T0245477

CAMBRIDGE LIBRARY COLLECTION

Books of enduring scholarly value

Earth Sciences

In the nineteenth century, geology emerged as a distinct academic discipline. It pointed the way towards the theory of evolution, as scientists including Gideon Mantell, Adam Sedgwick, Charles Lyell and Roderick Murchison began to use the evidence of minerals, rock formations and fossils to demonstrate that the earth was older by millions of years than the conventional, Bible-based wisdom had supposed. They argued convincingly that the climate, flora and fauna of the distant past could be deduced from geological evidence. Volcanic activity, the formation of mountains, and the action of glaciers and rivers, tides and ocean currents also became better understood. This series includes landmark publications by pioneers of the modern earth sciences, who advanced the scientific understanding of our planet and the processes by which it is constantly re-shaped.

Critical Examination of the First Principles of Geology

Born in London, the geologist G. B. Greenough FRS (1778–1855) initially studied law. His studies took him to the University of Göttingen where, almost by chance, he attended lectures on natural history. He was immediately hooked, gave up his legal studies, and devoted himself to geology, going on a series of scientific tours of France, Italy, Britain, Ireland and lastly India. He helped to found the Geological Society, and under its auspices, he organised a cooperative project that led to his famous geological map of England and Wales. He was made a Fellow of the Royal Society in 1807 for his services to geology. This influential series of essays, published in 1819, debunked a range of geological theories that were popular at the time, and by so doing, Greenough helped to reform much of geological thinking. The book also includes transcripts from his presidential addresses to the Geological Society.

Cambridge University Press has long been a pioneer in the reissuing of out-of-print titles from its own backlist, producing digital reprints of books that are still sought after by scholars and students but could not be reprinted economically using traditional technology. The Cambridge Library Collection extends this activity to a wider range of books which are still of importance to researchers and professionals, either for the source material they contain, or as landmarks in the history of their academic discipline.

Drawing from the world-renowned collections in the Cambridge University Library, and guided by the advice of experts in each subject area, Cambridge University Press is using state-of-the-art scanning machines in its own Printing House to capture the content of each book selected for inclusion. The files are processed to give a consistently clear, crisp image, and the books finished to the high quality standard for which the Press is recognised around the world. The latest print-on-demand technology ensures that the books will remain available indefinitely, and that orders for single or multiple copies can quickly be supplied.

The Cambridge Library Collection will bring back to life books of enduring scholarly value (including out-of-copyright works originally issued by other publishers) across a wide range of disciplines in the humanities and social sciences and in science and technology.

Critical Examination of the First Principles of Geology

In a Series of Essays

GEORGE BELLAS GREENOUGH

CAMBRIDGE
UNIVERSITY PRESS

CAMBRIDGE UNIVERSITY PRESS

Cambridge, New York, Melbourne, Madrid, Cape Town,
Singapore, São Paolo, Delhi, Tokyo, Mexico City

Published in the United States of America by Cambridge University Press, New York

www.cambridge.org
Information on this title: www.cambridge.org/9781108035323

© in this compilation Cambridge University Press 2011

This edition first published 1819
This digitally printed version 2011

ISBN 978-1-108-03532-3 Paperback

A

CRITICAL EXAMINATION

OF

THE FIRST PRINCIPLES

OF

GEOLOGY;

IN

A SERIES OF ESSAYS.

By G. B. GREENOUGH,

PRESIDENT OF THE GEOLOGICAL SOCIETY,

F.R.S. F.L.S.

LONDON:

Printed by Strahan and Spottiswoode, Printers-Street;
FOR LONGMAN, HURST, REES, ORME, AND BROWN,
PATERNOSTER-ROW.

1819.

PREFACE.

THE object of the following work is too clearly stated in the title page to require explanation. In furtherance of that object I have sometimes felt myself called upon to criticize the opinions of other writers. On those occasions I have done so with freedom, but I trust also without offence; for it were absurd to suppose that I do not feel respect for the intellectual and moral character of many persons, whose ideas on abstruse and speculative questions are at variance with mine. Errors emanating from men of acknowledged merit most require exposure, because they are the most contagious.

It should be recollected also, that many of the opinions here combated were advanced at a period when much less was known than is known at present, and would now, perhaps, if opportunity offered, be disclaimed even by their authors. I make this observation not in candour merely, but in prudence ; being satisfied that, if geological science continues to advance at the rate it has done lately, the essays now submitted to the public will, before many years have elapsed, be found to contain as many errors as they presume to correct.

CONTENTS.

ESSAYS

ON THE

ELEMENTS OF GEOLOGY.

ESSAY I.

ON STRATIFICATION.

STRATUM is a word so familiar to our ears
that it requires some degree of manliness
to acknowledge ourselves ignorant of its
meaning: the sense in which it is used is,
however, very far from being precise. Easy
as it may seem to determine whether a
given mass be, or be not stratified, there is,
perhaps, in the whole range of geological
investigation, no subject more pregnant
with controversy.

B

2

By way of illustration, I ask, whether granite is stratified?

" It is stratified, certainly," says [a] Gruber, " in the Riesengebirge; but do not pin " your faith on authority; take the evi-" dence of your senses; consult your eyes; " look at the rocks on the banks of the " Elbe, the Schneegrube, the Alpengrube, " the Schneekoppe; stratification is so " evident at all these places that no man " in his senses can doubt it."

" It is no less evident," says [b] Charpentier, " at the drey Steinen: and again at " Morgenstern," says Professor [c] Schubert, " at St. Gunther, at the Rathhausberg in " the forest of Bohemia, at Toplitz, at " Carlsbad."

" I can vouch for its stratification at Carls-" bad," says Deluc [d], " and if you want other " localities, you may add, on my authority,

[a] Gruber's Riesengebirge, p. 189.
[b] Beytrag zur Geogn. Kentniss. des Riesengeberges, p. 32.   [c] Geognosie, p. 119.
[c] Travels in France, Switzerland, and Germany, § 506, 525, § 549, § 743, &c. &c.

3

" Grosse Rad, Friesenstein, and the valleys
" of the Zackel and the Queis."

But why so particular?

" It is stratified," says Dr. ª Mitchell,
" along the chain of the Riesengebirge for
" fifty miles together." " For one hundred
" and fifty," says Professor Jameson.[b]

Yet M. von Buch [c] followed this chain
for nearly a hundred miles without being
able to discern, in any part of it, the
slightest trace of stratification.

Let us go from the Riesen to the
Erzgebirge.

" You will surely admit," says Professor
Schubert[d], " the granite at Johan-Georgen-
" stadt and Schwarzenberg to be stratified.
" M. von Buch[e] will not admit it."

[a] Scottish Isles, p. 37.
[b] Nicholson's Journal, vol. ii. p. 227.
[c] Journ. de Physique, vol. xlix. p. 211.
[d] Geognosie, p. 119
[e] Journ. de Physique vol. xlix. p. 211.

But the Fichtelgebirge.

The stratification of granite at the Ochsenkopf is recorded by M. de Luc. Mr. Buckland, Mr. Wm. Conybeare, and myself looked for stratification there, but in vain.

———

Saussure imagined, for some time, that Mont Blanc [b] was unstratified, but at length corrected his opinion. The mistake arose from its strata being of a thickness so enormous, that there are few points of view from which they are visible.

M. von Buch [c] says, that the granite of Mont Blanc is distinctly stratified; that the strata are vertical, or dip very gently to N. having the same direction as the chain.

M. Brochant affirms, that in the high Alps, granite is at times stratified, and that very decidedly. Gastern [d], in the neighbourhood of Salzburg, is cited by Schubert as affording stratified granite. At Pommat,

[a] Travels, vol. ii. p. 503.
[b] Voyages aux Alpes, § 919.
[c] Uber die Ursacken der Verbreitung grosser Alpengeschiebe in Tr. of R. A. of Berlin. 1815.
[d] Geogn. p. 119.

according to Ebel [a], the peasants slate their houses with it. At St. Roch [b], a hill on the Italian side of the Alps, Saussure tells us, " that there are eight distinct beds of " granite, the bottom one sixty feet thick, " the next fifty, the third twenty, the re- " maining five, forty, twenty-four, forty, " ten, forty. The lowest contains a half- " inch stratum of white feldspar, which is " parallel to the other strata. Similar " appearances present themselves between " Cresciano and Giornico."

Messrs. Brieslak and Isembardi [c] could discover no stratification in the granite north of Lario, and von Buch [d] denies that it is ever stratified in the Alps. " The " whole rock," he says, " is an assemblage " of crystals, united by the force of crystal- " lization," which was the opinion also of La Metherie.

Schubert [e] affirms, that granite is strati-fied in the Pyrenees. Cordier [f] admits this,

[a] Manuel. Pommat.
[b] Voyages aux Alpes, § 1752, § 1798.
[c] Brieslak's Geol. p. 156.
[d] Journal de Physique, vol. xlix. p. 211.
[e] Geogn. p. 119.   [f] Journal des Mines, vol. xvi.

6

adding, " that the strata exhibit no regula-
" rity of direction :" " such is precisely
" the case," says Dolomieu ª, " of all the
" granite which has fallen under my ob-
" servation."

To come nearer home. — Professor Jameson [b] states, that the stratification is very
distinct at Goatfield, the principal mountain in the Isle of Arran: Professor Playfair [c] is of the contrary opinion.

Professor Playfair [d] admits, that the granite of Mount Sorrel is stratified, not however without considerable hesitation, as his
companion, Lord Webb Seymour, does not
concur with him.

In the bed of the Ockment [e], in Devonshire, granite lies in slabs like lias.

M. de Luc [f] found stratification very evi-
dent in Cornwall at St. Columb, Tregonning
Hill, the Land's End, Castle Trereen; and
Mr. Conybeare at Cliggar. At St. Just the

1

[a] Journ. de Physique.   [b] Scottish Isles, p. 27 & 35.
[c] Illustrations, p. 329.
[d] Illustrations, p. 328.
[e] Geol. Trans. vol. iv. p. 402.
[f] Travels in England, § 1120, § 1168, § 1181, § 1189.

granite rises in large flag-stones. At St. Michael's Mount it exhibits parallel planes separated by layers, or wayboards of quartz. At Carglaze the parallelism is strikingly regular.

Dr. Berger, however, did not find granite stratified in Devonshire or Cornwall; nor has he found it so in any part of the Continent.

Dr. Hutton could never observe any regular structure in granite except that of divisions to which the shrinking of the mass gave rise after its consolidation.

Humboldt[a] discerns every where, in granite, strata, which in all parts of the world have the same dip and direction.

In one passage of Bergman's[b] works, mention is made of granite stratified horizontally. In another, the author says, " I " cannot allow any granite to be stratified " till it shall be found incumbent upon " some other rock."

---

[a] Journ. de Physique, vol. liii. p. 46.
[b] Geog. Physique, vol. i. Bergman.

Few authors [a] have expressed their opinion on the subject of syenite, porphyry, hornblend rock, greenstone, serpentine, primitive and transition limestone, siliceous slate, greywacke; but the stratification of these and other substances is no less ambiguous than that of granite. M. de Luc, for a long time denied, that any of the primitive, or as he calls them, primordial rocks were to be found stratified. This opinion was afterwards renounced by him [b], but still found a supporter in la Metherie. [c]

The term flötz-rocks was originally, I believe, employed upon the supposition that in them only was stratification observable, a supposition which is doubly erroneous, a stratiform structure, in whatever sense that phrase is used, being neither peculiar to flötz-rocks nor essential to them. Trap, a substance which derived its name from the circumstance of its occurring in successive steps, or terraces, piled one above another,

[a] Jameson's Geol. and J. des M. vol. xxvi. p. 173.
[b] Journ. de Physique, vol. xxxviii. p. 372.
[c] Theorie de la Terre, vol. iv. p. 352.

is denied, by Huttonian writers, all claim to the title of stratiform.

Whence this contrariety of opinion? Are our senses at variance, or our judgments? The cause I think is obvious. Every one uses the word stratum, no one enquires its meaning; the remedy is as obvious — definition.

Let us pass in review the several appearances in which this confusion has originated.

1st. Some rocks present external planes parallel to each other, but are not subdivided by internal planes.

Stratum is a literal translation of the word bed, and most writers use one or other of these expressions indifferently. Professor Jameson[a] not considering how injudicious it is to employ synonymes for the purpose of expressing contrast, has

[a] Geology, p. 51.

introduced a distinction between them.
Similar contiguous masses are by him
denominated strata, dissimilar ones beds.
Mr. Martin [a] has protested against this
innovation, and few authors without the
Wernerian pale appear disposed to adopt
it. Those who feel the value of such a
distinction, will do well, therefore, to select
some happier phrase to express it.

2d. Parallel internal planes are not always
co-extensive with the rock in which they
occur. Professor Jameson [b] observes, that
" the seams of the strata, or those lines
" which are said, in all cases, to mark the
" boundaries of a stratum, do not always
" continue throughout the whole mountain
" range." Brieslak affirms, that, " the
" strata of flötz rocks are often false, con-
" fined to the surface, an effect of decom-
" position; so that on one side you see
" them disposed in one direction, on the
" other, in a contrary direction. Attempt-

[a] Outlines, p. 170.
[b] Mem. of the Wern. Soc. vol. ii. pt. i. p. 222.
[c] Brieslak's Introd. à la Geologie, p. 233.

" ing to follow what seemed to be the part-
" ings of strata, he has found repeatedly,
" that it was impossible to do so to any
" distance." Dr. Pacard remarks [a], that,
" what appear to be beds, particularly in
" mountainous tracts, are often not so in
" reality, the appearance being given by
" fissures, veins, furrows, zones, folds, &c. ;
" that the interior of rocks is often very
" different from their exterior ; that he has
" seen these lines more or less superficial,
" suddenly stopped by a large block ; that
" the quarry-men often find zones, veins,
" &c. die away as they advance in the ex-
" cavation." Townson [b], speaking of some
rocks which he examined in Hungary, says,
" they were formed of one bed of compact
" limestone, about twenty-six yards thick.
" In one place it was pretty regularly divid-
" ed into four or five beds ; but these divi-
" sions, or signs of stratification, only ex-
" tended a few yards : between these beds
" were four or five thin beds of black silex
" running parallel to each other, but these

[a] Journ. de Physique, vol. xviii. p. 185.
[b] Townson's Travels in Hungary, p. 349.

" likewise were only of a few yards extent,
" and sometimes interrupted by the lime-
" stone."

" In some places," says Williams[a], " the
" mountain limestone appears regularly
" stratified, in others, one vast irregular bed
" of very unequal thickness."

3d. Parallel planes are sometimes too
distant from each other, at other times too
near each other, to give to the rock in which
they occur an undisputed claim to the title
of stratiform.

The strata of Mont Blanc, according to
Saussure[b], though horizontal, are not more
than three or four in number, the height of
that mountain being between fourteen and
fifteen thousand feet.

Fissile and slaty beds necessarily exhibit
parallel planes, yet many geologists do not
allow them to be stratified. [c]

[a] Mineral Kingdom, 2d ed. vol. i. p. 422.
[b] Voyages aux Alpes.
[c] " Although the slaty structure points out to us the
" direction which the strata must have, it does not follow
" that a rock having a slaty structure is stratified." —
Jameson's Min. vol. iii. p. 54. See also Journ. des M.
vol. xxvi. p. 173.

The slates in the canton of Glaris, interesting to naturalists on account of the fishes with which they are impressed, have been observed to be alternately thick and thin : M. Wittenbach, at Berne, threw out to me a very ingenious hypothesis to explain this singular fact : he supposed the thick slates formed by the ebb-tide, and the thin by the flood : in his opinion, then these slates could be no other than strata.

4th. Two or more sets of parallel planes sometimes occur in the same rock, so as to meet perpendicularly or obliquely.

At Ilfracombe, nodules of limestone occur in lines which form nearly a right angle with the planes of the slate in which they are contained : in the same neighbourhood may be seen parallel interrupted veins of quartz disposed in echellon, which also cross the planes of the strata. At Hedington Quarries, near Oxford, and at Anthony Hill, near Bath, the laminæ of the freestone are unconformable to the larger divisions. Mr. Townsend[a] mentions an appearance of the

[a] Townsend's Moses, p. 200.

same kind at Burnt-house Gate, and at
Swanwick, on the road from Bath to Glo-
cester; at Buckland Point, in the parish of
Mells, he found rhomboidal beds of oolite
truncated, perfectly smooth in the superior
and inferior surface, dipping at an angle of
40°, and confined between two horizontal
beds of clay.

At Westow, five miles from New Malton,
may be seen horizontal beds of oolite resting on highly inclined ones. To those who
wish other examples of this phenomenon, I
recommend an examination of the red
sandstone of Bridgenorth, or the Pennant
stone of the neighbourhood of Bristol.

Dr. Pacard[a], describing lines which cross
each other, one set being vertical, the other
horizontal, says, that this phenomenon may
be seen, among other places, in the gypsum
quarries at Mont Martre. [b] La Metherie
tells us, that granite is in general split in all
directions; that the fissures, when parallel,
may at first sight be confounded with seams

[a] Journ. de Physique, vol. xviii. p. 185.
[b] La Metherie, Theorie de la Terre, vol. iv. p. 352.

of stratification; but on looking more attentively we perceive they are not so.

Breislak [a], alluding to a remark of Pallas, says, that though many rocks of granite give one the idea of strata several feet thick, they are only rents which divide the rock into large parallelopipedons, and are to be viewed in the same light as the articulations of basalt.

One thing I must mention, says [b] Gruber, as occasioning me great perplexity, a perplexity which long experience alone would enable me to overcome: the rhomboidal clefts that occur in primitive districts and observe a mutual parallelism. At Aeskerbe the Elbe falls over a natural trellis formed by these clefts. At the Schnee-grube, and other mountains in the neighbourhood, you would suppose what you saw in the distance to be columnar basalt; on a nearer approach, however, it proves to be a large grained granite with parallel and vertical fissures.

[a] Breislak Geol. p. 156.
[b] Gruber's Riesengeb. p. 191. 4to.

At Luce Hill, a remarkable quarry, five miles from Hereford, the trap exhibits parallel fissures and stripes intersecting each other: similar appearances present themselves in the limestone rocks near Tenby, in Pembrokeshire, and in the slate rocks at Plymouth. At a projecting crag of greywacke, near Howth, on the shore of Dublin Bay, the position of the strata would seem to vary with that of the spectator, as in the hall of Greenwich Hospital, the eyes of a portrait on the ceiling appear to follow him to every part of the room. In slate rocks I know not of any criterion by which the planes of cleavage can be distinguished from those of stratification.

5. In rocks that have the appearance of deposition, the planes are not always parallel.

This happens by the very terms of the proposition in regard to all those strata which are said to be mantle-shaped, saddle-shaped [a], shield-shaped.

[a] Combles à deux croupes — demi combles — combles à croupes inegales ou λ. See l'Encycl. Geog. Phys. vol. iii. p. 442.

Mr. Jameson has observed, that we
sometimes find seams of several strata ter-
minating in the substance of a larger
stratum, and this in the substance of a
stratum still larger. The red sandstone of
Bridgenorth, Bromsgrove, Newton Ardes,
near Belfast, &c. is admirably calculated
for the study of this phenomenon.

In mountains of granite, where no other
parallelism is observable, it often happens
that the crystals of feldspar[a], or mica[b], lie in
the same direction.

5. The thickness of masses[c] varies considerably in different parts of their course.
It was the opinion of Buffon[d], that

There is a fine example of shield-shaped stratification
on the road from Freyberg to Dresden. The finest I
am acquainted with in Britain is at Coolnacarton, near
Ballynahinch, in the county of Galway.

[a] Deluc's Travels.

[b] Dolomieu Journ. des M. vol. vii. p. 426.

[c] I employ the word masses to express the *Gebirgs-
arten* of the Germans, not venturing, as some writers
have done, to apply the word Rock, associated from time
immemorial, with the idea of hardness and solidity to
the contents of a sand or clay pit.

Buffon, v. i. p. 172.

every stratum, whether horizontal or inclined, was of an equal thickness throughout its whole extent; and so it should be in
Wernerian[a] language by the terms of the
definition. Mr. Playfair[b], adopting this
idea, considers it a striking peculiarity of
the toadstone of Derbyshire, that even
when there is a thick covering of strata
over it, it has been found, by sinking perpendicular shafts, to vary from the thickness of eighteen yards to more than sixty,
within the horizontal distance of less than
a furlong. He insists on the cuneiform
shape which the rock at Salisbury Crag
takes at its extremities, and the great difference of thickness at them and in the
middle, as an argument extremely favourable to the Huttonian hypothesis: he
affirms that nothing of this kind is ever
found to take place in those beds of rock
which are certainly known to originate from
aqueous deposition, and that no character
can more strongly mark an essential difference of formation.

[a] Jameson's Mineralogy, vol. iii. p. 51.
[b] Illustrations, p. 294.

Mr. Farey[a], on the other hand, considers the wedge-like form a distinctive character of alluvial matters; these, he tells
us, never appear stratified in the uniform
manner peculiar to regular strata, but the
beds, when sunk through, frequently feather
out, or are wedge-like, and intermix with
each other.

Where the same property is represented
as characteristic of substances so different,
the presumption is that it will not be found
characteristic of any.

In the Memoir of Cuvier[b] and Brongniart on the environs of Paris, it is stated,
that before the later formations, mentioned
in that memoir, were deposited, the chalk
upon which they rest must have presented
a surface varied with depressions and protuberances. These inequalities are clearly
ascertained by the islands and promontories
of chalk which occasionally penetrate the

[a] Farey's Survey of Derbyshire p. 131. See also his
Remarks on the Stratification of France and England,
in Philosophical Magazine.
[b] Recherches sur les Ossemens Fossiles, tom. i. p. 14
and 15.

incumbent beds, and by the very uncertain
depth at which that substance is found in
wells and excavations. It is also stated,
that the thickness of the plastic clay varies
from sixteen metres and more, to one or two
decimetres.

The same irregularity is described by
Williams [a] as occurring in strong beds of
the mountain limestone. — " In many
" places," he says, " I have seen the lime-
" rock swell out and increase from two or
" three fathoms to more than a hundred
" fathoms thick."

In another [b] part of his work, this Author makes a similar observation in regard to coal. " The same individual
" seam," he says, " seldom preserves the
" same thickness to the extent of four or
" five miles on the line of bearing; I have
" seen some so variable, that you could
" not depend on finding the coal equally
" thick for twenty yards together in any
" part of the field."

Some striking instances of the wedge-

[a] Min. Kingd. edit. 2. vol. i. p. 403.   See also p. 57.
[b] Ibid. p. 60.

like form of the stone-bands, which occur in the coal-mines of Newcastle, are recorded by Mr. Winch in the Transactions of the Geological Society.[a] The same phenomena may be conveniently observed in the beds of sandstone, limestone, and slate, at Bangor Ferry.

" If," says Mr. Aikin [b], " we confine our " attention to those beds which lie between " the big and little *flints*, and which con-" stitute by far the most regular part of the " great coal-field of Shropshire, we shall " find that the *pennystone* bed, which, in " the Madeley pit, varies in thickness from " six to eight feet, is fifteen feet thick at " Lightmoor, about twenty feet at Dawley, " sixteen feet at Old Park, and eighteen " feet at Ketley; that the Viger coal, with " its superincumbent clay, occupies a " thickness of about twelve feet at Made-" ley, is diminished to three feet at Light-" moor, and is entirely wanting in all the " collieries which lie to the north of the

[a] Geological Transactions, vol. iv. p. 12.
[b] Ibid. vol. i. p. 198.

" latter; that a bed of clay, usually known
" by the name of the *Upper Clunches,*
" bears a thickness of from fifteen to
" twenty-six feet in all the above-men-
" tioned collieries, except that of Ketley,
" where it is entirely wanting."

I am told by a person on whose authority
I can rely, that near Ashby de la Zouche,
the *bend,* separating the second and third
seam of coal, is, in the easternmost coal-
pits, thirty-three yards thick; in the next,
to the west, twenty-five; in the most western
only fourteen; and that in the Bedworth
collieries, about half a mile further to the
west than these, it vanishes entirely, the
second and third coal-seam running together.

Robinson[a], generalizing this observation,
ascribes the wedge-like shape to all coal
whatever. " As in all classes of coal the
" seams gradually increase in thickness
" till they come to their full height and
" growth, so they gradually decrease till

[a] Essay towards a Nat. Hist. of Westmoreland and
Cumberland, p. 49.

" they dwindle out into small seams, and
" then the covers change and the coal goes
" out."

Dr. Holland[a] informs us, that in Cheshire
there seems to be a progressive thinning
of the upper bed of salt from north-west
to south-east. " In the mines sunk near
" the west or north-west of the salt district,
" the thickness of this bed has been gene-
" rally twenty-eight, twenty-nine, or thirty
" yards; proceeding towards the east, or
" south-east, it decreases to twenty-five, and
" towards the eastern boundary to twenty,
" eighteen, and seventeen yards."

Prof. Jameson[b] says, " that the sandstone
" and lime-rock, in the river district of the
" Forth, afford many examples of this kind."

" However we may be struck," says Von
Buch[c], " with the circumstance that the
" porphyry under the zircon syenit, at
" Gretsen, does not, perhaps, reach a hun-
" dred feet in thickness, and rises within

[a] Geological Transactions, vol. i. p. 43.
[b] Von Buch's Norway, p. 67.   Note in Mr. Black's
Translation.
[c] Ibid. p. 66.

c 4

" four English miles from thence to the
" height of thirteen hundred feet and up-
" wards, such examples are not altogether
" new in geognosy; opposite Ringerige, in
" the same neighbourhood the sandstone
" beneath this porphry is above eight
" hundred feet thick ; but at Gretsen and
" at Giellebeck it entirely fails."

Mr. Jameson observes, " that great in-
" equality in the thickness of beds of por-
" phyry and sandstone is a common ap-
" pearance in Great Britain."

The *liegende Stocke* of the [a] Germans
(lying masses, if literally translated, but
which I would rather distinguish by the
name of lenticular masses), are defined to
be portions of earthy or metallic matter,
in form somewhat resembling a wool-sack,
contemporaneous, but not co-extensive,
with the beds in which they lie: their
ordinary length and breadth may be from
twenty to fifty fathom ; their thickness
from ten to twenty ; they taper off in both
directions, from the middle towards the
extremities. Rock-salt, and the ores of

[a] Jameson's Mineralogy, vol. iii. p. 226.

iron and copper, are particularly remarked as occurring in masses of this description.

According to Mr. Jameson, these lenticular masses are rarely seen in primitive
but more frequently in flötz mountains:
otherwise I should have said that granite,
sienit, hornblend-rock, and greenstone-rock
were more usually found under these circumstances than any other.

6. The term Stratification is by no means
unconnected with theory. To constitute a
Stratum it is not enough, in the opinion of
many geologists, that a substance divides
in parallel planes, unless it does so in consequence of the manner in which its particles were at first deposited from a fluid
menstruum; now it is obvious that as long
as the propriety of using the term depends
on the idea we entertain of the manner in
which strata were formed, unity of opinion
can alone preclude confusion of language.
Let us enquire, therefore, how far geologists agree in their opinions upon this
subject.

It was the opinion of Woodward, and, I believe, of all other geologists who wrote during the last century, that strata were the effects of deposition alternately suspended and renewed; that the loose materials from which they were formed subsiding at the bottom of the sea, and naturally yielding on the side opposite to that where the pressure was greatest, had arranged themselves in horizontal layers, the vibration of the incumbent fluid, by impressing a slight motion backward and forward on the materials of these layers, naturally assisting the accuracy of their level.

In this mode is supposed to have been formed the first horizontal stratum. New materials, furnished at successive periods, would obey the same law, and the surfaces of each stratum being parallel to that of the water from which it was deposited, would be parallel to that of every other stratum.

That an interval of time, greater or smaller, elapsed between the several depositions, seems ascertained. At Ingleton, limestone is separated from the rock beneath it by the intervention of loose pebbles, perfectly agreeing in substance with the rock they cover. The oolite at Marquise, near Boulogne, reposes on a bed of marble, the upper surface of which is furrowed by water and perforated by vermiculi.

To these arguments may be added another furnished by the frequency of wayboards, of coral reefs, of beds of shells, and the extraordinary circumstance, now so much insisted upon, of shells peculiar to fresh-water alternately covering and being covered by other shells, which belong only to the sea.

If it is difficult to conceive the existence of so large a quantity of water as would be necessary, under present circumstances, to hold the entire globe in solution ; that difficulty, if not removed, is at least much diminished, by supposing the different parts to have been in solution at different periods.

Again, it has been argued, that had the deposition proceeded uninterruptedly, the particles deposited, instead of being arranged as we now find them, must have followed each other in the order of their specific gravity. An argument particularly applicable to the cases of beds of brecchia and sandstone.

It seems highly probable, therefore, that the parallel planes which the surface of different beds exhibit, are, in many cases, the effects of deposition alternately suspended and renewed; but that they are not so in every case may be inferred, I think, with equal confidence, from the following considerations : —

1. Parallelism of surface alone does not prove that the rock in which it is observable has been deposited by water; for parallelism may be produced by other causes, as every one admits, in basaltic pillars, in backs and cutters, in the laminæ of crystals. In beds of fibrous gypsum, tabular flint, &c., though the external surfaces are parallel, it is evident from an examination of the interior structure, that these beds have been formed, not by the

superposition of successive particles, but by crystalline shoots from the upper and lower surfaces towards the middle. The cheeks of mineral veins, or dykes, are sometimes parallel for a considerable distance, though they cannot be supposed to have derived this parallelism from a disposition to conform to the horizontal surface of a fluid.

2. The more or less frequent recurrence of parallel planes depends on the nature of the substance deposited. Granite, porphyry, serpentine, trap-rock, salt, chalk, are seldom, if ever, found in other than thick masses. Argillite as constantly in flakes: sandstone and oolite in beds of moderate thickness. It is a remark of Bergman [a], that granular rocks are seldom stratified. Williams observed [b], that brecchia appeared stratified, or not, in proportion to the size of the component parts. Townson [c] made the same observation at Grau, in Hungary. Where the brecchia is

[a] Journ. des M. vol. iii. p. 66.　[b] Mineral Kingdom.
[c] Travels in Hungary, p. 62.

very coarse, nothing like stratification, he says, is to be perceived; but in the same hill the brecchia is often as fine as a sandstone, and is then more or less distinctly stratified. He adds, that the same thing happens in these kingdoms, and particularly in the neighbourhood of Edinburgh.

3. The larger divisions of rocks are frequently not parallel to the lamina, of which these rocks are composed.

4. In regard to way-boards they seem to depend no less on the nature of the adjoining rocks, than on the circumstances which attended their formation. Stony beds generally have them, but sands, clays, loams, are divided from one another only by a change of colour or very imperfect suture; and in those substances which are called primitive, no way-boards have yet been discovered.

Catcott has given a philosophical explanation of this phenomenon. " If you take " a certain portion of earthy bodies, and

" pulverize them to the finest degree ima-
" ginable, and mix them as confusedly
" together as possible, and let them fall
" through a dry fluid, such as the air, they
" will settle just in the same confused state
" as they were; if you permit them to sub-
" side through water, they will settle more
" or less in parallel strata. Indeed it re-
" quires twenty or thirty times the quantity
" of water to earth to make this layer-like
" subsidence tolerably apparent even in
" the mixture of but three or four bodies;
" but the greater quantity of water you
" use, and the finer you pulverize the sub-
" stances, the more apparent and regular
" the strata will be." (P. 269.)

According to Werner, strata are from
four to six feet apart in the older formations, but less distant in the newer.

Hutchinson observes [a], " that in the mid-
" land counties of England the strata are
" commonly thin near the surface, and be-
" come gradually thicker in proportion to

[a] Hutchinson's Works, vol. xii. p. 264.

" their depth ; but no thin strata whatever
" are found in Cornwall."

ᵃ " At La Porte de France, one of the
" quarries at Grenoble, the lower beds are
" thin and laminated, rather slate than
" limestone ; higher up they are thicker
" and more calcareous. In the middle are
" procured large blocks of marble. Above
" these beds are others of common lime-
" stone, and the remaining ones, as they are
" successively softer, less calcareous, more
" argillaceous, as they approach nearer the
" surface, successively diminish in thick-
" ness."

Townsend remarks, that no soil such as
covers the present surface of the earth is
found between the strata.

5. At the junction of two kinds of rock,
as greywacke slate, and limestone, sandstone, and trap, chalk and green sand, we
often find that each is impregnated with
the substance of the other.

6. The contemperaneous veins of one

stratum sometimes extend themselves into the adjoining stratum. At Mouzainville, a village between Varennes and Verdun, M. Ferrusac[a] observed a bed of grey limestone, six feet thick, lying horizontally between beds of clay: in some places the two external beds were traversed by veins issuing from the third: in others they graduated into each other, and amalgamated: at an adjacent quarry they again appeared distinct, but the intermediate one was greatly changed in quality and thickness: presently this intermediate one blended with its neighbours, sent out a fork, ramified, was lost: yet the total thickness of the three beds continued uniformly the same.

Cuvier, carried away by natural partiality to the subjects which have most engaged his attention, is, perhaps, a little unjust towards those which have escaped it. According to him, neither way-boards, nor pebbles of known rocks imbedded in other rocks, nor horizontal strata resting upon inclined ones, could ever have led to the

[a] Journal de Physique, vol. xv. p. 453.

supposition that one part of our globe
was older than another. " Is it not clear,"
he exclaims, " that we are indebted for
" a theory of the earth to fossils only?
" Without them, who would have dreamt
" of the globe having been formed at
" successive epochs, and by a series of dif-
" ferent operations? It is only by analogy
" that we extend to primitive districts the
" conclusions which fossils enable us to
" form in regard to secondary districts;
" if all strata had been without fossils, no
" one could maintain that they were not
" all formed at the same period." [a]

7. Decomposition, or torrefaction, will
often expose to view stratification previously latent; the slate of the Cotswolds
has no appearance of slate till acted upon
by the frost; and for a similar reason the
upper beds of every quarry are in general
thinner than those beneath them.

8. " Depositions which go on uninter-

[a] Cuvier's Recherches sur les Ossemens Fossiles,
tom. i. p. 35.

" ruptedly in our laboratories arrange them-
" selves in distinct layers."

———

Such are the arguments which induce me
to think with M. Ferrusac[a], and Mr. Jameson [b], that contiguous strata may in some
instances be contemporaneous.

———

The various postures which rocks assume give rise also to much diversity of
opinion.

The position of masses is determined by
the direction and inclination, or the dip [c]
and inclination, observed through their
whole extent.

Their Direction is the position with regard to the meridian of an imaginary

[a] Journal de Physique, vol. xv. p. 454.
[b] Wernerian Transactions, vol. ii. p. 225.
[c] For the Wernerian definition of these terms, see
Journal des Mines, vol. xxvi. p. 170.

straight line, formed by the intersection of their planes with that of the horizon.

Their Dip is the position with regard to the meridian of an imaginary horizontal line drawn at right angles to the line of direction.

Their Inclination is the measure of the angle formed by the intersection of their planes with that of the horizon.

If the direction is given, the dip is determined, and if the dip is given, the direction is determined.[a]

The instruments required for determining the position of a mass, are a compass and clinometer.

The Clinometer presented to the Geological Society by Lord Webb Seymour, would be perfect, if sufficiently portable. A smaller one, constructed by Mr. Jaffray, has the advantage of enabling us to determine the dip by measuring any two sections of a stratum, provided they are

[a] Pini was one of the first writers who felt the importance of recording the dip and direction of beds, (see his Observations on St. Gothard); practical miners, however, must have attended to this subject much earlier.

not parallel to each other. With the common clinometer, when the planes of a stratum are inaccessible, the sections measured must be adjacent.

All that a clinometer will teach us, however, is the posture of a mass in the precise spot where it is examined; but the direction and dip may be very different in one part of a country from what they are in another: thus in Kent, the chalk dips to the North, in Sussex to the South, in Wiltshire to the East, in part of Hampshire to the West. This instrument is, therefore, by no means adapted to the use of those whose object it is to discover the average of dip.

In using the compass, we should bear in mind that many rocks are magnetic; and in recording our observations with this instrument, should state whether we have allowed for its variation.

In Germany, the miners' compass is graduated as a dial, divided into twenty-four parts, or hours: by those who employ it, a stratum dipping to the East is said to dip to six o'clock.

In the scientific language of this country, it is usual to express the inclination of a bed by the angle it makes with the horizon; but the expression among labouring people is, it dips so many inches in the yard.

In travelling over an extent of country, the direction of the beds is characterized by sameness and uniformity. Every slope is opposed to another of corresponding steepness, unless when other causes operate so as to render insensible the operation of this. There is little variety in the productions of the land, or the condition and employment of the inhabitants. In travelling along the line of dip, on the contrary, our eyes are continually regaled with a change of scenery. Every hill has a character of its own, and is succeeded either by hills of a different character, or not unfrequently by a flat. Where the ascent is steep on one side, the descent is gradual on the other. Commons succeed to inclosures, pasture to arable, wolds to marshes, a

12

naked district to a forest, a poor soil to one
remarkable for its fertility, and *vice versa.*
The valleys extend principally to the right
and left, rarely showing themselves in front.
The employment of the people is various,
and the fences and buildings are constructed of different materials.

In crossing strata obliquely, these characters are combined.

THE POSITIONS OF MASSES.

The positions of masses and strata have
been classed by the writers of the Encyclopedie [a] under the following heads : 1. parallel to the horizon ; 2. perpendicular ;
3. inclined [b] ; 4. curved inwards ; 5. curved
outwards ; 6. curved upwards ; 7. curved

[a] Encyclopedie Geog. Physique, vol. i. p. 32.

[b] Mr. Pinkerton recommends that we should forego
the use of the term strata, as applied to inclined masses,
substituting the term Arrects; but if this recommendation were adopted, I fear we should lose more in sound
than we should gain in sense.

downwards ; 8. circular ; 9. undulating ; 10. zigzag.

It has been supposed, that the primitive rocks were always vertically, the secondary horizontally, stratified. It is true, that vertical planes occur more frequently in the older rocks than in the newer; but it is also true, that every rock, in different parts of its course, exhibits both these appearances.

At St. Roch, on the south side of the Alps, and near Cresciano, Granite, according to Saussure [a], is horizontally stratified. Deluc, in his letters to Professor Blumenbach, mentions another instance at Missouri. Dr. Mitchell [b] and Mr. Jameson observed the same thing at the Riesengebirge. Charpentier [c] tells us, of Granite in Saxony which is horizontal. Bergman [d] also speaks of horizontal Granite, though

[a] Voyages, § 1752.
[b] Nicholson's Journal, vol. ii. p. 227.
[c] Charpentier's Mineralogische Geographie der Chursachsischen Lande, p. 17 and 389.
[d] Journal des Mines, vol. iii. p. 75.

without specifying the exact place at which he had observed granite in this position.

Saussure [a] found Gneiss horizontal at Monte Rosa. I have seen it lying horizontally upon a bed of feldspar-porphyry, near Nieder Schonau, on the road from Freyberg to Dresden, and Charpentier tells us that this is its usual position. In the neighbourhood of Freyberg, it is commonly thought that the Gneiss is more horizontal than the mica-slate, and the mica-slate than the clay-slate. Between Mohorn and Herzogswald, the clay-slate appeared to me to form an angle with the horizon of between 20° and 30°, the mica-slate between 15° and 20°.

Horizontal beds of Hornstone-slate have been observed by Bergman [b] at several places on the confines of Jemteland and Norway.

M. Picot de la Peyrouse [c] says, " that ho-

[a] Voyages, § 2138.

[b] Bergman Physicalische Beschreibung der Erdkugel, 3d edition, vol. i. p. 192. vol. iii. p. 75.

[c] Journal des Mines, vol. vii. p. 51.

" rizontal strata of primitive Limestone al-
" ternate in one part of the Pyrenees, with
" horizontal strata of Corneenne. The beds
" of Limestone which occur in Sienite, be-
" tween Meissen and Dresden, are also hori-
" zontal."

The Slate of Wales and Cumberland
is in some places horizontal.[a]

On the other hand, there is perhaps no
rock which does not occasionally appear in
strata, inclined at a considerable angle to
the horizon.

It has been observed, that the secondary
Sandstones and Limestones [a] of the Pyrenees, are often vertical, and that a dip
of less than 45° is a circumstance almost
as rare at the base of these mountains as
at the summit, at the extremities as in the
centre of the chain.

The high inclination of the Sandstone of
the Vallorsine, so much insisted upon by
Saussure, is analogous to what we find at

[a] Journal des Mines, vol. xii. p. 88.

the Lecky in Worcestershire, at Alderley-
edge in Cheshire, in the district of Gower,
and the Ridgeway in Pembrokeshire. At
Callendar, in Scotland, the old Conglomerate
lies in beds absolutely vertical.

The Transition [a], or Mountain Limestone,
is nearly vertical at Caldey Island, at Tenby,
Dunraven, the Mumbles, the Clee hills,
&c. &c.

The Magnesian Limestone at Bredon, in
Leicestershire, forms with the horizon an
angle of at least 45°.

The Coal measures of this Country are
often inclined at a very considerable angle;
those at Talbenny, in Pembrokeshire, are
nearly vertical.

At Brancilli [b], near La Claitte, in Flanders, their dip varies from 60° to 70°, and
at the hills of St. Giles [c], near Liege, they

[a] The mountain Limestone is the transition Limestone of Werner, not, as is usually supposed, the first
flötz.

[b] La Metherie, Theorie de la Terre, vol. v. p. 59. and
vol. iv. p. 153.

[c] l'Encycl. Geographie Physique, vol. i. p. 709.

assume every position from the vertical to the horizontal.

The Red Marl has a very considerable dip [a] on the east side of Catton in Croxall.

Near Little Stoke, in Somersetshire, the Lias [b] is vertical. I have seen it vertical also at Eckersberg, north-east of Jena, where the disturbance has been so great, that in one place the red marl would seem to rest upon it.

Vertical or highly inclined beds of Chalk occur at Handfast Point on the coast of Dorsetshire, at the Hog's back in Surrey, the Needles in the Isle of Wight, &c.

At Alum Bay, the Clays and Sands that cover the chalk are vertical.

It has been erroneously imagined by some writers, that all rocks are horizontal in plains, and inclined in mountains.

---

[a] Tilloch's Magazine, March 1810, p. 136.
[b] Geological Transactions, vol. iii. p. 373.

Was the inclination of masses and strata given to them at the time of their deposition, or has it been the result of subsequent catastrophes?

The following arguments have been adduced to prove the strata originally horizontal.

1. It is not unusual to observe upon the opposite sides of a fault the same strata following each other in the same order, but having their planes on the one side parallel to the horizon, on the other inclined : there can be little doubt that these were originally continuous, and that one of them has shifted its position. Now, it is infinitely more reasonable to suppose, that the disturbance which occasioned this change, has thrown the strata out of their horizontal bearing than into it. This argument is therefore conclusive in all cases to which it applies.

2. In some cases an original inclination of the strata seems incompatible with the

nature of their materials. Can it be believed, that the unconsolidated sand and marl of Alum bay, now vertical, has been vertical from the beginning? In the neighbourhood of Bristol, two series of beds, similar in character, but differing materially in respect of age, occur in very different positions ; the strata of old red are nearly vertical, those of the newer red nearly horizontal ; is it not highly probable that they were in the first instance both horizontal?

" It is impossible," says Mr. Aikin [a], speaking of a rock in Shropshire, " that a bed " of sandstone, and much more of clay, " marl, or mud, as it no doubt was in its " original state, should have been dispos- " ed on a plane, at an elevation of from " 30° to 40°, in such a manner as to con- " stitute an extensive stratum of a uniform " thickness, and that hardly exceeding a " foot for a depth of at least a hundred " feet."

". The Tubulites, which pierce the " marl at Steeraway, lead to the same " conclusion : some of these, twelve inches

[a] Geological Transactions, vol. i. p. 206.

" in length, and scarcely the eighth of an inch
" in diameter, being perpendicular to the
" planes of the stratum, are inclined at an
" angle of 50° to the horizon, now it is the
" known habit of these animals to affect
" a vertical position, with regard to the
" horizon."

The manner in which the pebbles and
bowlder-stones occur in the older brecchias, renders it impossible that beds of
this description should have possessed
originally their present inclination. Where
pebbles [a] lie in veins, as in the Relistian
mine, they are disposed horizontally in
the order of their specific gravities; whereas, in these conglomerates, they are disposed in that which would have been the
order of their respective gravities, if the
beds had been horizontal; consequently
they could not remain an instant in their
present position if the cement which unites
them were to become soft.

This argument first suggested itself to M.
Saussure on examining the celebrated pud-

[a] Philosophical Transactions, see also Werner on
Veins, p. 61.

ding stone of the Valorsine : the importance which Dolomieu attached to it will
be seen by the following passage.

" This consequence[a] (viz. that rocks
" were originally horizontal) has been con-
" firmed by the beautiful observations of
" M. de Saussure. He it was who, by find-
" ing beds of rolled stones, first established
" the fact, the most important there is in
" Geology, that the beds on which these
" stones lie were once horizontal, and both
" subsequently thrown out of their position
" together. The shifting of these beds,
" then, may be placed among those funda-
" mental truths which serve as a founda-
" tion for all systems. I have long felt
" the necessity of admitting it, but for M.
" de Saussure was reserved the honour
" of proving it. He has done more for the
" advancement of Geology than all the na-
" turalists who preceded him."

M. Deluc acquiesces in the panegyric
just quoted ; after recording M. de Saussure's
change of opinion, (for at first he supposed

[a] Journal de Physique, vol. xxxix. p. 384.

inclined strata to have been always inclined,) and his motive in altering it;— " thus," he says, " has he diffused upon Geology a " light which can never be ob- " scured; whoever shall attentively study " the nature of mountains, as I have " done since I have had that light to " guide me, in a variety of countries, will " every where find the same phenomena, " which through the eye irresistibly speak " to the understanding."

Werner [a] noticed a conglomerate in the coal-field of Hainichen, composed almost entirely of flat pieces of clay-slate of considerable size, having in some places a position nearly vertical, the same with that of the strata : it is impossible, he observes, that these stones should have been arranged in this manner by the action of water ; they must therefore have taken their position afterwards along with the strata. The pebbles contained in the inclined

[a] Werner's Theory of the Formation of Veins, translated by C. Anderson, M.D. p. 60.

E

" strata" of sandstone at Nottingham have the flat side downwards.

3. Sir James Hall [a], describing the inclined and contorted strata near St. Abb's head in Berwickshire, remarks, that many of the beds possess that peculiar undulation on their surface which we meet with on a sandy beach when the tide has left it, and which affords the most unequivocal indication of aqueous deposition. If so, the inclination of these beds cannot be of the same date as themselves.

4. Now, if it be admitted, that in the instances above-mentioned, the present position of the beds is not their original position, it seems extremely difficult not to admit more; for the inclination of these is often the same as the inclination of the beds immediately beneath,

[a] Edinburgh Transactions, vol. vii. p. 82. In this memoir, the author supposes the killas (as he calls it, I think improperly) to have shifted its position since the red sandstone was deposited.

so that it appears very improbable that any one should have shifted its place, while the others retained it : of this, Saussure has given instances in Switzerland, and our own country affords many; in Wales, and more strikingly perhaps in the south of Ireland, the conglomerate rock is so intimately connected with the sandstone and slate of transition (if I may use that term), and these again with the older rocks, that it is hard to say where the series begins or terminates. No catastrophe can be supposed to have affected one without affecting the whole.

The interposition of pudding-stone, says Deluc [a], begins even among the primordial strata. Saussure has seen it between granites and shisti, between shisti and calcareous strata, between calcareous strata and sandstone.

These are the principal arguments that occur to me on the one side : let us enumerate those which may be brought forward on the other.

---

[a] Deluc on Geology, p. 374.

1. It is evident from the nature of veins, that under favorable circumstances the particles of matter may arrange themselves in beds highly inclined to the horizon. We obtain the same conclusion from a circumstance already noticed, that the laminæ of rocks are often inclined where the strata themselves are horizontal. Within a mile of Cernioge, the stripes are at right angles to the laminæ of the slate on which they occur : here then is another instance of correct parallelism, without the aid of horizontal deposition.

2. Inclined strata are frequently unaccompanied by any mark of violence, and exhibit the greatest regularity both in form and direction.

Whatever difficulties may be involved in the contrary hypothesis, M. Cordier [a] is startled at the introduction of a force which could displace the entire chain of the Pyrenees at once, and yet be so regular in its direction as to leave traces of its effect in a straight line only.

[a] Cordier, Journal des Mines, vol. xvi. p. 280.

Picot de la Peyrouse[a], describing the
same mountains tells us, that the regularity,
the sportiveness, the caprice of the vertical
beds of sand and limestone, heterogeneous
nodules of one rock in the heart of another, the uniform verticality of the beds
which compose the primitive ridges and
secondary summits, vertical beds traversed
by horizontal ones, preclude the idea of
any sudden or irregular force having shifted
the rocks of the Pyrenees into their actual
posture from one originally horizontal.

Humboldt[b] has observed " strata" of
clay-slate dipping N.W. at an angle of
70° for fifteen successive leagues. Can
it be imagined that these strata were
once horizontal, and consequently fifteen
leagues thick? or that falling, they found
a chasm large enough to receive them?
But he goes further, and maintains this
to be the common inclination of all primitive strata throughout the globe. His
argument would be strong, if the facts

[a] Journal des Mines, vol. vii. p. 65.
[b] Journal de Physique, vol. liii. p. 47.

on which he grounds it were well established.

3. Beds[a] which are vertical, or inclined at a considerable angle to the horizon, are, in many cases, broader at the base than they are at the summit, or, in other words, mantle-shaped. This form can hardly have been given to them by any change of posture subsequent to their consolidation.

4. M. von Buch[b] derives an argument in favour of the original inclination of rocks from the change which is often observed in their nature on the opposite sides of a granitic chain : thus, on the Italian side of the Alps, and between Bolzano and Brixen, in the Tyrol, porphyry is very abundant, constituting eminences four thousand feet above the level of the sea ; but this rock fails altogether on the side of Germany and Switzerland. On the northern side, mag-

[a] " Beds which are perpendicular at the surface of " the earth become gradually more horizontal as they " descend." Jameson's Mineralogy, 1st edit. vol. iii. p. 224.

[b] Journal de Physique, vol. xlix. p. 212.

nesian rocks abound, which are extremely rare on the southern. If these rocks derived their inclined posture from elevation or subsidence, why are they not found on both sides of the chain?

Secondary Limestone occurs very abundantly on the north-eastern flank of the primitive chains in Saxony, and Bohemia, the Alps, the Carpacks, and even in North America ; but very sparingly on the south-eastern flank of those chains.

The coal-measures on the opposite sides of the Derbyshire limestone are dissimilar, and the shisti on the opposite sides of the Ocrinian chain.

Humboldt [a] remarks, that the direction of high chains of mountains seems to have great influence on the direction of strata, even at a considerable distance. Of this influence which Humboldt has observed in the Pyrenees and in Mexico Ebel [b] has discovered numerous instances in the Alps.

5. The irregularities observed in the stratification of some beds seem to warrant the

[a] Journal de Physique, vol. lxxi. p. 372.
[b] Ebel, vol. i. p. 220. vcl. ii. p. 201. 215. and 357.

opinion that these beds were originally inclined : thus at Malvern [a] the inclination
of the sandstone diminishes as it recedes
from the hill, and then increases ; to this
succeed strata of limestone, the inclination
of which becomes more and more considerable ; these are followed by strata of an
argillaceous rock which continue to rise
till they become vertical, and then dip westward with a gradually-diminishing angle
of inclination.

6. Secondary rocks are generally inclined
at their junction with the primary.

7. Secondary rocks are often unconformable to the primitive rocks, on which
they rest ; in Red Bay, Anglesea, the former dip to S.E., the latter to N.W.

At the head of the lac de Joux, in the
Pays de Vaux, vertical beds are said to rest
on horizontal ones ; perhaps fissure has
been confounded with stratification here,
as at the Saleve, where the same incon-

Geological Transactions, vol. i. p. 306. See also La
Metherie, Theorie de la Terre, vol. v. p. 81. where he
describes the Coal-measures near Liege.

gruity has been supposed to prevail between the strata at the base of the hill, and those higher up.[a]

If the beds were originally horizontal, and afterwards shifted, the supposable causes of this shift are, 1. An internal force acting from below upwards, so as to raise the crust of the globe. 2. A want of support, owing to internal cavities, so that the beds have fallen by their own gravity. 3. An external shock which has broken the shell, and made one part tumble over another. Dolomieu[b] was inclined to the last of these opinions.

But to what circumstances can we ascribe this inclination of rocks, if we suppose it original?

[a] The Huttonian doctrine upon this subject is curious. Assuming that all rocks have been forced up once from the bottom of the sea, and primitive rocks twice, it supposes that the primitive rocks have acquired their vertical or inclined posture in consequence of this violent elevation, the secondary in spight of it. — *Playfair's Illustrations*, p. 123.

[b] Journal de Physique, vol. xxxix. p. 385.

1. Mr. Hutchinson[a], who supposes slate
and killas to have been deposited mechani-
cally, attributes the inclination of their beds
to the shape of their component particles.
Bodies of such a flaky form, he conceives,
would subside edgeways. In coal-measures, however, we know that slate is found
in a posture nearly horizontal.

2. Many authors[b] ascribe the original inclination of strata to an inequality in the
base or ground on which they were deposited. The figure of the lower beds deposited on an uneven surface would necessarily be affected, says Mr. Playfair[c],
by two causes, the inclination of that surface, on the one hand, and the tendency to
horizontality on the other; but as the former cause would grow less powerful as
the distance from the bottom increased, the
latter cause would finally prevail, so that
the upper beds would approach to horizontality, and the lower would neither be

[a] Hutchinson's Works, vol. xii.

[b] La Metherie, Theorie de la Terre, vol. v. p. 55.
Jameson's Mineralogy, 1st edit. vol. iii. p. 55.

[c] Illustrations, p. 43.

exactly parallel to them, nor to one another. In a small work published by the famous enthusiast Swedenberg[a] in 1722, the following experiment is recorded:

" Mud and water, well mixed together,
" being poured into a vessel, the bottom
" of which was conical, subsided hori-
" zontally : the form of the bottom did
" not produce any effect. More mud and
" water being thrown in, the strata de-
" posited were still horizontal ; but the
" mixture being allowed to stand a day
" or two, the strata during this time
" shrunk considerably towards the edge
" of the vessel, and very little towards the
" centre ; the degree of consolidation
" varied as the depth, and the strata
" gradually became inclined."

Where this cause has operated, therefore, we ought to find the strata in different parts of their course unequal in thickness and solidity : Does not this happen in mantle-shaped strata ?

Dr. Pacard[b] made a series of experiments, with the same view. He found that

[a] Miscellanea, pars i. p. 23.
[b] Journal de Physique, vol. xviii. p. 189.

a quantity of earth or salt suspended in a liquid adhered to the sides of the vessel when inclined, but when upright, was parallel to the surface of the water; and inferred that the obliquity of strata depended wholly on the circumstances under which the deposition took place. May not corpuscular attraction, that agent by which our tea-kettles are coated and our decanters furred, have operated on a more enlarged scale in the formation of veins and strata?

3. Professor Werner[a] was the first, as far as I know, who considered the primitive rocks as composed purely of crystalline matter, without denying them, however, the attributes of stratification: the transition, or those which he supposes to have been produced by the joint agency of crystallization and gravity, are in his system also admitted to be stratified.

In what manner all the parts can be dis-

[a] See an attack on this Theory in the Edinburgh Review, vol. ii. p. 343. and Kirwan's Answer. See also l'Encyl. Geog. Physique, vol. iii. p. 504.

posed according to the laws of crystalline aggregation, while the whole is obedient to those of gravity, is a question natural to ask, and not easy to resolve.

The effects attributed to crystallization by later writers are considerably more extensive. La Metherie could not fail to refer to crystalline agency the divisions of strata and the construction of mountains, since the human form itself appeared to him to be the necessary result of that process. Mr. Jameson is of opinion that strata are analogous to the laminæ of crystals, and viewed on the large scale, would probably be found to meet each other at determinate angles.

All masses and strata are subject to curvature and angularity.

The curvatures of Gneiss are particularly specified by Mr. Jameson [a] as characteriz-

* Jameson's Mineralogy, 1st edit. vol. iii. and Trans. of Wern. Nat. Hist. Soc. vol. i. p. 290.

ing one of the members of that formation ;
those of Mica-slate are very conspicuous
at Ben Lawers, Ben Lomond, and Glen-
croe ; and what would be remarkable if
it were true, it is said that garnets are
never found in this description of rock, unless where it is waved. The promontory of
Holyhead exhibits to great advantage the
inflexions of a Chlorite-slate, in which
quartz is predominant ; inflexions of Killas
and Transition-slate are seen on the coast
of Cornwall[a], and Devon ; at Aberystwith ;
in the neighbourhood of Llangollen; at
Conway ; at Llawrwyst bridge; and on the
road from thence to Corwen — of Primitive
Marble at Derrowra in Conemara.

The mountain chains of the Continent
abound in similar examples.

M. Duhamel [b], speaking of the Pyrenees,
says, " compact feldspar (roche de corne),
" trap, limestone, form an immense mass,
" composed of a prodigious number of al-
" ternate beds of no great thickness, and in-

[a] See Geological Transactions, vol. ii. and the Port-
folios of Geol. Soc.

[b] Journal des Mines, vol. viii. p. 753. and again p. 750.

" clined to the horizon at a high angle.
' Some of them are plain and regular, while
" others twisted in a thousand different
" directions, without disturbing the paral-
" lelism of the beds above or beneath, give
" you the idea of an explosion." Contortions in the Old Red Sandstone are not
common in this country; but they are
seen at Jedburgh; fine examples occur in
the Pyrenees [a] at Port Vieil and la fourche
d'Allans. — Near Eckersberg, in Saxony,
the Red Ground containing fibrous gypsum
is slightly contorted.

At Lucan, and on the shore towards Por-
train, north of Dublin, the curvatures of
Calp are exceedingly complicated. At
Mercaston in Derbyshire, at Chepstow, at
Orme's Head, we observe them in Mountain
Limestone ; — at Bredon hill in Magnesian
Limestone ; — at Ashford [b] in Limestone
Shale ; — at Sandsfoot [c] near Tenby, on the

[a] Journal des Mines, vol. vii. p. 54.
[b] Several other instances are given in the Survey of
Derbyshire, vol. i. p. 231.
[c] See drawing in possession of the Geol. Soc.

banks of the Dee between Holywell and Mostyn; within a mile of Berwick upon Tweed; and still more strikingly at Liddel Brig near Castleton in Roxburgshire in Coal measures; — at Dunraven and Barry Island in Lias.

The more modern rocks which occur at Purbeck and along the coast of Dorsetshire are often contorted. A very remarkable contorsion of Chalk at Handfast Point, is figured in the work of Sir Henry Englefield. Dr. Pacard[a] describes a similar inflexion at Mont Martre.

At Port Dinnleyn in Carnarvonshire, I found the same waving lines in sandstone now forming, in the Dunes on the sea shore; and a very fine sand alternating with Coaly Matter, (both of them being alluvial,) forms fan-like or wedge-shaped strata, on the west side of the road which leads from Edinburgh to Leith.

[a] Journal de Physique, vol. xviii. p. 185.

Examples of striking contorsions in secondary rocks, on the Continent, may be seen in Journal des Mines, vol. xviii. p. 307, and vol. ix. p. 449.

I cannot therefore admit with Mr. Allan[a], that contorsions are characteristic of transition slate ; nor subscribe to the more common opinion, that they are peculiar to primary strata, or if found in more recent rocks, found only on the verge of the primary.

The circle, of which these curves are the segments, are sometimes many miles, sometimes only a few inches in diameter.

Mr. Playfair [b] mentions with astonishment, the smallness of some of the curves ; other authors have been equally struck with their vastness. In some places, says Williams [c], they are on so large a scale, that the waves of water do not convey an adequate idea of the spacious dimensions of the troughs, and of the extent and magnitude of the ridges. Hutchinson [d] has made use of the same figure. I have never seen the surface of the sea more variously curled in

[a] Allan on Transition Rocks, Edin. Transactions, vol. vii. p. 90.   [b] Illustrations, p. 221.
[c] Mineral Kingdom, 2d. edit. vol. i. p. 90.
[d] Hutchinson's Works, vol. xii.

a great storm, than I have seen the grain of the same stratum of killas.

Saussure[a] has compared strata which he observed in various parts of the Alps, to the letters Z S C to a disjointed X, to the Greek Lambda λ, &c.

———

It is supposed by Mr. Playfair[b], that the curvature is generally, if not universally, simple, like the superficies of a cylinder, not double like that of a sphere ; — this is a mistake.

As an instance of curvature extending in both directions, we may mention mantle-shaped strata. This appearance, though it has been most observed in primitive rocks, is by no means peculiar to these ; in the north of England the limestone mantles round the slate ; the coal-measures of Derbyshire mantle round the limestone.

When masses or strata decline upon every side towards a certain point, they are said to be basin-shaped. Such is the disposition

[a] Saussure, Voyages aux Alpes, § 36. 467. 475.
[b] Illustrations, p. 226.

of the mountain limestone at Ormeshead, of the coal in South Wales, of the chalk in the north of Ireland.

The clam-shell cave at Staffa was probably so named, from the conchoidal form which it derives from curvature in the strata. It is supposed that on the great Clee hill in Shropshire, there are no less than seven distinct coal-fields ; the principal of them is covered by basalt, which varies in thickness from 60 yards to 0, though this coal-field is only two miles in length, and one and a half in breadth : the strata dipping to a common centre, the thickness diminishes towards the circumference.

Another coal-field, a quarter of a mile in diameter, situate on the same hill, crops out in both directions.

In all these cases, the curvature is plainly not cylindrical but spherical.

Let us now consider the probable causes of curvature.

1. According to Dr. Hutton, the strata

having been formed horizontally, were lifted, while flexible and ductile, in a direction from below, upwards ; owing to gravity, and the resistance of the mass, this direction became oblique, and the lateral force occasioned contorsions.

Such is Mr. Playfair's[a] statement, and he adds, that the Huttonian theory is no where stronger, than in what relates to the elevation and inflexion of the strata ; points in which all others are so egregiously defective. [b] It is with surprize, therefore, that I find one of the most zealous supporters of that theory, allowing us an alternative on this occasion : Sir James Hall[c], supposes that the strata, originally horizontal, have been urged, when in a soft but tough and ductile state, by a powerful force, acting horizontally, and opposed by an insurmountable resistance on the opposite sides ; or, that the same effect has been produced by two forces acting in opposite directions, at the same time that the whole was held down by a superincumbent weight,

[a] Illustrations, p. 45.　　　[b] Ibid. p. 234.
[c] Trans. of R. S. E. vol. vii. part i. p. 84.

which, however, was capable of being heaved up by a sufficiently powerful exertion.

In regard to Mr. Playfair's supposition :

1. I do not understand how crystalline strata could be so lifted, while flexible and ductile, because in regard to these there is no middle stage between fluidity and consolidation.

2. From the operation of such a cause would result other indications of disturbance which have not resulted.

3. In many cases no such cause can have operated, for the curved strata rest on horizontal strata in which there is no curvature. The nagelfluh of the Rigi is stratified with the utmost regularity ; the limestone which covers it is extremely disturbed and contorted: how happened it that a force which so affected the limestone, elevated from 5 to 7000 feet above the level of the sea, did not affect the nagelfluh beneath ?

The limestones of Dudley, Aberly, Long-hope, Malvern, &c. the shale of Ashford, the calp of Dublin, are all remarkably contorted, but none of these rest upon contorted strata.

On the coast between Scarborough and Filey, curved strata lie on strata that are not curved; there is a similar instance at Tenby, in South Wales.

4. If we admit the strata to have been elevated, it is easy to imagine that two opposite forces acting upon a body, the stronger upwards, the weaker downwards, should communicate to that body a lateral motion ; but will such motion account for the phenomena in question ? would it not rather close the fissures which in many cases are not closed? consolidate the mass, which in many cases is not consolidated? if it incurvated the strata, would the curves be cylindrical, or in one direction only? or would they not rather spread like a bush, upwards, North, South, East, and West, forming a species of curve unknown in the history of stratification ?

5. If the theory of Mr. Playfair were true, curvature should never be found in those strata that are horizontal, whereas horizontal strata are often curved.

6. The conformity of different strata is another circumstance fatal to this hypothesis. There is no species of rock in which the curves are more frequent or more fantastical than greywacke slate ; this rock, we know, in many instances, alternates with conglomerate, the pebbles of which are disposed in such a manner, that it would be impossible for them to remain an instant in the place which they occupy, if the cement which connects them together were to become soft. The conglomerate, therefore, and consequently the slate which alternates with it, could not have been elevated, till after its consolidation. If, then, as the Huttonians say, it was not consolidated till after it was curved ; neither was it elevated till after it was curved: in other words, the effect preceded the cause.

In regard to the supposition of Sir James

Hall, it seems to me that the particles being exposed to horizontal pressure (how produced we are not informed, and will not stop to enquire) would ease away to the part where resistance was least. If it were least towards the top, the mass would bulge upwards; but will this explain the complicated phenomena of curvature?

M. de Saussure, speaking of strata, which he found on the road to St. Martin, disposed in the form of a divided X, observes, " These curves are so exactly " true throughout, and so continuous, " that I can never believe the strata " have been horizontal in the first in- " stance, and afterwards thrown into the " fanciful position in which we now be- " hold them. Not only must the mass " have been soft when acted upon by the " disturbing cause, (for there is no appear- " ance of fracture in the most sharp and " sudden inflections,) but it must have been " treated with a gentleness and delicacy " which it is impossible to describe. It is " not in the nature of violent convulsion

" to preserve so carefully the continuity of
" the parts." [a]

At the hill on which Dudley Castle is
built, the interior strata are very nearly vertical: the exterior are folded round them, and
their edges join, after making the circuit of
the hill, so as to form a ring. Wren's
Nest, another eminence in the neighbourhood of Dudley, is constructed of the same
substances, arranged in the same manner.

II. In the Wernerian [b] theory the contorsions of some of the primitive masses are
explained upon the principle of crystallization.

This idea, first suggested by Saussure[c],
is condemned by Mr. Playfair[d], as unsatisfac-
tory and illusive. The purpose for which
crystallization is here introduced, he says,
is not to give a specific figure to a particular substance, but to derange the substances, which it has formed and figured
according to certain rules, a work which

[a] Voyages aux Alpes, vol. ii. § 475.
[b] Jameson's Mineralogy 1st edit. vol. iii. p. 352.
[c] Voyages aux Alpes, § 159. 475.
[d] Illustrations p. 232.

we know not how it is to perform, and in which we have no experience of its power. Mr. Murray[a] attempts to get over this objection by an appeal to facts ; and pointing out the analogy which subsists between the contorsions of entire strata, and the wavings which, on the small scale are seen in alabaster and stalactite, (wavings which Count Bournon[b] ascribes to a change in the direction of the constituent crystals,) infers that as crystallization has produced these appearances in the one case, it may be presumed to have done so in the other. I apprehend it has not produced them in either, and that in reasoning on this subject Mr. Murray has not guarded himself sufficiently against an error to which we are always liable, that of confounding co-existent phenomena with cause and effect. That alternate beds of different substances, such as limestone, sandstone, and slate, should be all curved, as we find them on the coast of Devonshire, and in many

[a] Comparative View of the Huttonian and Neptunian Systems of Geology, p. 109.

[b] Bournon's Traité de la Chaux Carbonateé, vol. i. p. 187.

other places, seems fatal to the theory which would ascribe these curves to crystallization.

The curvature of masses may perhaps depend, sometimes on an unequal effect, produced by temperature on the materials which compose them; sometimes on the motion of the fluid from which they were deposited; sometimes on the form of the bottom on which they rest.

III. On unequal effect produced by temperature on the materials;
The curvatures which I refer to this head, are analogous to the warping of wood, the curling of parchment, the blistering of paint. It is well known, that trees are flattened in beds of clay or marl. At Newcastle-on-Tyne, a portion of coal being removed, the remainder is allowed to stand for the purpose of supporting the roof; in a few years, however, this support is no longer necessary. The shale bulges out in the interstices, and a new set of pillars now do the

office which those of coal had performed
before ; a fresh supply of coal is obtained ;
the interstices are again filled up by the
bulging of the shale, and the workings are
thus suspended and renewed till the mine
is finally abandoned.

I am informed that horse adits have been
tried at the Great Clee Hill, in Shropshire,
and failed there simply in consequence of
the swelling of the argillaceous beds,
which takes place to so great an extent
that a party of men is employed every
night to enlarge or prop up the adits in
which another party had worked during the
day : without this precaution, the galleries
would in a short time be closed.

Sometimes, says Williams [a], " the roof
" falls down for a certain space below its
" ordinary level, and squeezes the coal
" much thinner, especially in the middle
" of that space. These accidents are of
" various dimensions, some of them not
" more than two or three feet in diameter ;
" these are like a wart or protuberance on
" the under side of the roof of the coal, and

[a] Mineral Kingdom, 2d. edit. vol. i. p. 60.

" sink into the upper side of the coal, like
" the bottom of a great pot. Others are
" two or three yards in diameter, and some
" even thirty or forty. The large ones do
" not thrust themselves into the coal so
" abruptly as the small ones, but press down
" gradually with a gentle swell."

Where clay alternates with limestone or silex, may not contorsions have taken place in consequence of the unequal rate at which these substances would consolidate? and will not the contorsions which are the most sudden, be found in those substances, which at the time they became contorted, were the most ductile?

IV. On the motion of the fluid from which they were deposited.

In the mica slate which occurs between Lough Mask and Lough Corrib, in Galway, the mica has a determinate direction, and its laminæ are straight, but the intervening quarz is sometimes straight, sometimes curved, and varies so materially in its direction, that what in one spot would be supposed to be a conformable layer, exhibits in another all the characters of a vein.

Near Cherbourg, between the little cove of le Poulet and Becquet, a small village on the east of Bretteville, where beds of killas, both plane and curved, are striped by veins of quarz, I could not fail to observe that the curvatures were most sensible in those parts in which the quarz was most abundant.

Mr. Hutchinson [a], speaking of inflexions in killas, remarks that this substance was probably thrown into that irregular and unnatural posture by some irregular agitation of the water; in corroboration of which he states that the straight slates are finer and firmer than those which are inflected.

Mr. Calcot [b] pursues this idea further.

In the veins of some sorts of stone, he says, it is common to observe a great variety of matter in the greatest variety of forms and directions; in some part, matter that was lighter than the neighbouring, pressed down below the place due to its specific gravity, and afterwards elevated to a consi-

---

[a] Hutchinson's Works, vol. xii. p. 333.
[b] On the Deluge, p. 263.

derable height, till at last, meeting with
matter that was heavier, and making its
way downwards, the whole shall be curved
by the ascent of the one and the descent of
the other into a vast variety of arches, consisting of the finest and most delicate lines;
in other parts you may see streaks or seams
of different substances proceeding on, as it
were, horizontally, in nearly straight lines,
till they have been met and opposed by
other matter in a contrary direction, and at
the point of conflux both species of matter
turned back and deflected in all the variety
of wave-like dispositions that can well be
imagined to have happened in two streams
of water meeting each other in opposite
currents : in short you may see all the di-
versities of forms in the solid, that any kind
of agitation in a fluid could display.

But the author who has expressed himself most eloquently in favor of this opinion is M. Ramond. [a]

" Will any man attempt," says he, " to
" explain the disorder which prevails in the
" Pyrenees, by supposing the beds to have
" been regularly deposited in the first in-

[a] Journal de Physique, vol. 53. p. 139.

" stance, and shifted afterwards by sub-
" sidences, shocks, convulsions? Sure I
" am, that nature did not so operate at the
" Pic d'Eredlitz, where trap, petrosilex, &c.
" wind in and out among plane strata of
" lime-stone — no, nor at the Pic de Midi,
" where regular beds of massive granite,
" containing beds of limestone no less re-
" gular, are traversed by whimsical creepers
" of horn-stone, gneiss, and even granite,
" which have wandered away from the main
" body of that substance. By what mira-
" culous influence did the limestone of
" Eredlitz continue plane, amid the violent
" concussion by which the siliceous veins
" were injected into it? These creepers, if
" I may call them so, at the Pic du Midi,
" what external force could so contort them,
" naturally protected as they are between two
" walls of granite? And then the granite,
" which must have acquired all the hard-
" ness it has at the moment of its crystalliz-
" ation, how did it contrive to insinuate itself
" with impunity into the intermediate lime-
" stone and to wind about there with all the
" pliability and suppleness of wax? What
" convulsions, what subsidences marvellously

" combined, sparing the surface of the rocks
" to rage more furiously within them, gave
" to the constituent parts of the marble of
" Estaubé that rotatory motion to which we
" find nothing at all similar in the entire
" mass, save only a moderate curvature. Is a
" crash, or a blow, sufficient to explain, not
" only the undulations of the limestone of
" Sers, but also the veins that traverse it ?
" A stratum may get bent by sliding over
" another stratum — be it so : the play of
" crystallization, the accident of shrinking,
" may produce inflexions in heterogeneous
" rocks. I will not dispute even that ; but
" that either of these causes, or both of
" them, however modified, could pro-
" duce this great movement, preserving
" the harmony of the whole, yet spreading
" disorder through all the parts, is a doc-
" trine contradicted by the disposition and
" nature of the ingredients, by the structure
" of the masses, by a comparison of the
" facts, and by the aspect of the spot."

" Here are no beds which any one can
" suspect of having been once regularly
" horizontal, continuous, and of equal thick-

G

82

" ness throughout. The compact limes-
" tones, the semi-argillaceous marbles, and
" slates of the mountains of Estaubé, the
" sandy lime-stones, the brecchias and grits
" of Mont-Perdu, seem to have been dri-
" ven against one another by opposite
" forces, which, at the point of contact,
" have shivered them into short, irregular,
" tortuous veins, the entanglement of which
" forms the intervening masses.

" Figure to yourself a number of viscid
" liquors, differently coloured, spreading
" themselves in whirling laminæ, in the
" vessel into which they are poured; watch
" a thick column of smoak floating in the
" air; you will then have before you an
" image of the confusion which prevails in
" these rocks; perhaps an explanation of
" it. While the waters were depositing
" the secondary strata, impetuous currents
" from the south disturbed this process, by
" bringing in a quantity of mud, sand, and
" rubbish; the struggle of the two con-
" flicting masses, the repeated assaults of
" the one, the persevering resistance of
" the other,— this is the idea naturally sug-

2

" gested, on contemplating the contorted
" veins in these rocks ; 'tis a sea consoli-
" dated in a storm, the violence of which,
" may still be seen in the petrified waves."

The undulating furrows in sand, upon
the sea shore, are obviously occasioned by
the unequal advance of the waves ; the
incrustation upon the baths of Baiæ, exhibits similar wavings, derived from the same
cause. To this cause, perhaps, we may
attribute the wavings of alabaster, and of
the primitive shisti.

Dr. Pacard[a], who made a very interesting
series of experiments to explain the phenomena of stratification, shews, that curved
strata may be artificially produced, by
merely throwing into a tub of water, earths
of different degrees of fineness, taking care
that the last put in shall be the coarsest
and heaviest.

At Ashford[b], the strata of limestone
shale, are, in one place, flat and regular,

[a] Journal de Physique, vol. xviii. p. 187.
[b] Agricult. Survey of Derbyshire, vol. i. p. 331.

and at a short distance you find, without any apparent cause, the most contorted strata that can be seen.

The remarkable curve of some beds of chalk at Handfast-point, in the Isle of Purbeck, figured and described by Mr. Webster, in the work of Sir Henry Englefield, is an instance of the same kind.

Mr. Jameson[a] states, that clay-slate is sometimes waved even in veins.

V. The last cause which we have to consider, is a disposition on the part of the materials, to conform to the shape of the ground on which they are deposited ; a disposition of the cast, to take the form of the mould.

Mr. Playfair[b] supposes, that curved strata preserve a constant equality of thickness and distance among the component laminæ ; but we are informed, by the author of the Article, Argyleshire, in the

[a] Jameson's Mineralogy, edit. 1. vol. iii. p. 351.
[b] Illustr. p. 232.

Encyclopædia Britannica, that, in that
country where the strata consist chiefly of
limestone, with few or very thin strata
of slate intervening, the thickness of a
stratum is frequently five or six times
greater at the summit of the wave, and
at the hollow where it begins ascending
to form the next wave, than at the immediate point where the contrary flexure
takes place.

Such, I believe, will be found to be
the case almost invariably in masses or
strata, which are said, in Wernerian language, to be mantle-shaped, shield-shaped,
saddle-shaped, basin-shaped, trough-shaped ; I say almost invariably, because, in an
analogous instance, that of agates, the
force of attraction seems to have completely overpowered that of gravity, and
the thickness of the laminæ in different
places affords no criterion of their position ; it will happen almost invariably, in
these cases, that the masses will not be
respectively parallel, the lower surface of
each will follow the inequalities of the

mould, and the upper surface, in proportion as it is further removed from this, will become more and more horizontal.

Dr. Pacard [a] observed at the quarries at Belleville, that where one arch pointed upwards and another downwards, the strata, thus deprived of their parallelism, left an oval space, either open or filled, with extraneous matter.

In the neck of land situate between Lough Mask and Lough Corrib, the mica-slate is more contorted the nearer it approaches the greenstone.

In the valley of Ashover the limestone and superincumbent shale make a considerable sweep in order to surmount the toadstone, where it swells out beyond its usual proportions. At Crich, in the same county, where there is another protuberance of the toadstone, the " strata" of limestone rise with a moderate inclination, till it attains the western brow of the cliff, then become vertical for about 600 feet, and then nearly horizontal. Notwithstanding these sudden changes of position, no trace is to be ob served of fracture or dislocation.

At Dunraven, the strata of lias are folded in like manner over a rock of mountain limestone.

———

It is asked by Mr. Playfair if the curvature of the strata arose from the irregularity of the bottom on which they were deposited, why is it in one dimension only, and not in every direction, like that of hills or valleys, or the actual surface of the earth? Let us ascertain the fact before we endeavour to explain it. Williams, as if anticipating an objection not made till after his death, expressly tells us, that the curvatures which he describes, resemble those parallel ridges and vallies which diversify the face of a country.

According to Lehman[a], most of the secondary strata present hollows or depressions; those observed in the neighbour-

[a] Lehman's Flötz Gebirge, p. 137.

hood of Ellesmere have been supposed to
possess characters peculiar to the sandstone
in which they occur; and beds of gypsum
are particularly noticed by the Freyberg
school on account of the funnel-shaped
hollows that appear in them. Some naturalists affirm that all coal-measures are deposited in troughs and basins, and dip inwards to a common point.

Strata occupying basins I have generally
observed to be most disturbed where the
basins suddenly contract, as the coal-measures of Pembrokeshire. Professor Raumer's Memoir furnishes me with a similar
example at Flinsberg and Harrachsdorf in
Silesia.

To conclude, then, let me ask, Where a
rock is *stratified*, is it necessarily bounded
by parallel surfaces? if so, let us hear no
more of mantle-shaped, saddle-shaped,
shield-shaped, fan-shaped, basin-shaped,
trough-shaped stratification.

Are its surfaces necessarily parallel to those of the adjoining rock? if so, let us hear no more of unconformable and overlying stratification.

Is it sufficient that parallelism shall be found in a portion of the rock? Let us never hear of substances being unstratified. Or must it extend through the entire mass? Let us hear no more of strata.

The laminæ of flagstone, the folia of slate, are these strata? Are laminæ, four hundred yards thick, strata? Is there any assignable limit to their thickness or tenuity?

When one set of parallel planes crosses another, are both sets to be called strata, or neither, or only one of them? If one only, by what rule are we to be guided in distinguishing the real from the counterfeit?

Must the beds be so arranged as to convey to the observer the idea of deposition alternately suspended and renewed? If this is not necessary, how is the parallelism derived from stratification, to be distinguished from parallelism resulting from other causes? and of what use is it to know whether a substance is stratified or not? If it is necessary, where two observers have imbibed contrary impressions, how shall we determine which of the two is right?

Let him who can answer these questions rest assured that he has a distinct idea of stratification.

ESSAY II.

ON THE FIGURE OF THE EARTH.

THE figure which the earth actually presents, is called its real figure; its statical is that which it would present if level with the surface of the sea.

ON THE STATICAL FIGURE OF THE EARTH.

Let the earth be supposed fluid to a certain depth; then the statical figure which it would assume, in consequence of rotation upon its axis, would be that of a spheroid flattened at the poles.

Such, or nearly such, being the figure which it has assumed, we have good reason to believe from this circumstance alone, that the earth has been more or less fluid to a certain depth; and on examining the

substances of which its crust is composed,
we find in confirmation of that belief, evident marks that these substances have all
existed at some period or other in a soft or
fluid state, and most of them in a state of
aqueous solution or suspension. Whence
so large a supply of water can have been
obtained as would be necessary to render
fluid the entire surface of the globe, what
can have become of it when obtained, are
indeed questions far beyond the limit of our
knowledge, if not beyond that of rational
conjecture ; but in Geology, as in all other
sciences, it will happen continually, that we
are unable to discredit what we are unable
to explain. Before we yield or refuse as-
sent to any proposition, we must sum up
probabilities and improbabilities on both
sides, and strike a balance. Now in the
case before us, it is less extraordinary that
water should have stood in some former
period at a height exceeding that of our
highest mountains, than that crystals should
have been formed without a solvent, — strata
without a sediment or precipitate, — that
consolidation should have taken place with-

out the union of parts, — desiccation without moisture, — that fishes innumerable
should have lived without water, — and
gravel have been rounded without attrition.
Almost every writer, therefore, who has
touched upon this subject, admits the original fluidity of the earth ; — not so, however, the Huttonians ; they recognize in
nature no trace either of a beginning or an
end ; they know of nothing original.

Omnia mortali mutantur lege creata,
At manet incolumis mundus suaque omnia servat ;
Quæ nec longa dies auget, minuitve senectus,
Nec motus puncto currit, cursusque fatigat.
Idem semper erit, quoniam semper fuit idem :
Non alium videre patres, aliumve minores
Aspicient ; deus est qui non mutatur in ævum.

Manilius Astr. i. 515.

Accordingly an ingenious, but in my
opinion ineffectual attempt, has been made
by one of the most eminent writers of that
school, to deduce the statical figure of the
earth from the gradual changes which occur in its actual figure. These changes he
attributes to two causes ; one the continual
wear and tear of the surface ; the other the

production of new lands, volcanic or not
volcanic, thrown up from the bottom of the
sea at indefinite periods, to uncertain alti-
tudes, by heat how excited, how supported,
whence gifted on a sudden with explosive
power, we are not informed, and thrown up
always in the absence of witnesses. These
two causes are said to counterbalance each
other, and we are not prepared, therefore, to
anticipate from them any effect. But we are
told, that the general tendency to produce in
the earth a spheroidal figure, " may" still remain, and more ª, " may" be done by every
revolution, to bring about the attainment of
that figure, than to cause a deviation from
it: to that figure, therefore, the earth by a
mere change of tense, " will" continually
approach. — Continually approach ? then
in time it will acquire it. Mr. Playfair en-
tertains a different opinion ; upon what
founded he does not inform us, but contents
himself with observing, that the Huttonian
theory affords, what no theory had before
done, a satisfactory explanation of the statical figure of the earth.

` Illustrations, p. 504.

ON THE REAL FIGURE OF THE EARTH

The actual figure of the earth is diversified by eminences, depressions, and plains
of uncertain extent and elevation.

Compared with the solid contents of our
planet, these irregularities are insignificant,
of no more moment, to use the expressive language of Seneca, than particles of dust on the
surface of an artificial globe. It is always
useful to enlarge our ideas of nature by re-
flecting on the comparative littleness of those
objects which we are accustomed to consider the most sublime. The interest, however, which mountain and valley are calculated to excite in us, depending not upon
their relations to our planet, but upon their
relations to our species, is little affected by
any comparison that may be instituted between their magnitude and that of the world
at large.

From the internal evidence which the
complexion of our earth affords, I propose
to trace its inequalities of surface, first
to their proximate cause, afterwards to
those causes which are more remote.

ON THE PROXIMATE CAUSE OF THE INEQUA-
LITIES NOW SUBSISTING ON THE SURFACE
OF THE EARTH.

Some changes are now going on, and
ever have been since history began, from
recent volcanoes, coral reefs, dunes, and
calcareous concretions. These, however,
which are much less considerable in amount
than we are at first disposed to represent
them, have been so ably pourtrayed by M.
Cuvier, in his well known preliminary dis-
sertation, that it is unnecessary to do more
than declare my entire concurrence in the
view which he has taken of this part of the
subject.

The interstices between mountains and
hills, have been produced, for the most
part[a], by the removal of matter which formerly occupied them.

In countries where the strata are horizontal, opposite hills having the same elevation, generally consist of the same sub-

[a] The rule does not apply to volcanic countries, sand
dunes, and coral reefs.

stances; the beds found on the heights are wanting in the bottom of the valley; those which lie beneath these beds in geological order, are not wanting. Sometimes the lower half of a bed is found in the bottom of a valley, the upper half only on its side; followed to the head of a comb, the opposite strata approach, unite. This correspondence cannot escape the most careless observer in those vallies, the walls of which are precipitous, and at no great distance from each other. It often exists, however, though not so strikingly displayed, in wider vallies, where the opposite banks are separated by large rivers and arms of the sea.

On the opposite sides of the Thames, the gravel and sands of Blackheath correspond with those of Epping; and the clays of Hampstead, Highgate, and Harrow, with those of Richmond, Wimbledon, and Sydenham. In Derbyshire, and Yorkshire, we find numerous insulated summits of millstone grit, incumbent on a plain, consisting of mountain limestone. The chalk which prevails on both sides the English channel,

and extends to Flamborough Head, recurs
in the islands of Jutland, Zealand, and
Rugen. That of Dorsetshire re-appears in
the Isle of Wight, and in the same extraordinary posture. The Lias, at the mouth of
the Seine, is covered immediately by green
sand, to the exclusion of the various clays,
grits, and oolites, which generally divide
them. The same irregularity happens on
the opposite coast at Lyme, and at both
places are discovered the remains of fossil
crocodiles.

The hard white limestone and trap, found
on the north coast of Antrim, are precisely
the same as those of the opposite island of
Rathlin. The ridge of shistus which
crosses the south of Scotland, from St.
Abb's head to Portpatrick, re-appears at
Donaghadee, in Ireland.

The granite of Iona is found again at the
south-western extremity of Mull; and that
of the Land's End, in the islands of Scilly.
The sienite and superincumbent killas of
the Morne mountains in Ireland, are identical with those which compose the Lowren
and Criffel mountains in Scotland.

Buffon[a] informs us, " that the Maldiva
" islands, which, when taken together,
" extend about 200 leagues in length, are
" divided into thirteen clusters : each is
" surrounded with a chain of rocks of the
" same stone, and there are only three or
" four small and dangerous openings, by
" which they can be approached ; they are
" all placed in a line, with their ends to
" each other, and appear evidently to have
" been a long mountain crowned with rock."
The Bahama Islands all consist of the
same limestone as the nearest point of the
mainland, and the resemblance which M.
Herminier[b] has observed in the mineral
productions of the Antilles and those of
the adjacent continent, induces him to
suppose that these islands have been continuous, and the gulph of Mexico a Mediterranean sea.

The bed of the Adriatic[c] is of shell marble, the same as that of the mountains on
both sides. The strata of Syracuse correspond with those of Malta, and the strata

[a] Buffon, Smellie's Translation, vol. i. p. 177.
[b] Journal de Physique.
[c] Donati, Journal de Physique, vol. ii. p. 593.

of Malta with those of Gozo. The same striking resemblance in the products of opposite islands has been observed in the Grecian Archipelago.

What is the natural inference to be drawn from these several circumstances of agreement? " If a person were to see the broken " walls of a palace or castle that had been " in part demolished, he would trace the " lines in which the walls had been carried, " and in thought, fill up the breaches, and " reunite the whole. In the same man- " ner," says Mr. Catcott[a], " when we view " the naked ends or broken edges of strata " on one side of a valley, and compare " them with their correspondent ends on " the other, we cannot but perceive that " the intermediate space was once filled up, " and the strata continued from mountain " to mountain."

Strange, that some writers who admit the original continuity of all other rocks, however discontinuous we now find them, should so lose sight of analogy, when they

[a] Treatise on the Deluge, p. 163.

speak of basaltic rocks, as to imagine the occurrence of these in insulated hummocks, a remarkable phenomenon, out of the common course of nature, and to be explained only by the capricious interference of their favourite idol Vulcan or Pluto!

Another point in which this agreement may be observed, is in mineral veins. I know not of any instance in which the course of a vein or dyke has been cut off by a valley, so as not to recur on the opposite side.

It is observed by Dolomieu [a], that the same veins are distinguishable on either side of the valleys situate near Monte Rosa.

At Killarney a lead mine is worked under the lake : at Clontarf, near Dublin, beneath the Bay : at the Wherry Mine, near Penzance, and several others on the coast between Cape Cornwall and St. Ives, for tin and copper, the workings for tin and copper have been carried under the sea.

That the strata which constitute the coast of Antrim, once extended further than they

* Journal des Mines, tom. vii. p. 424.

do at present, is proved by the basaltic[a] dykes which project into the sea. Near Carrickfergus, many of the points or headlands, are formed by dykes of the same nature, which certainly cannot have always terminated where they do at present.

This correspondence may be observed also in the nature of the soil, gravel, and bowlder stones.

" At the back of the rocks in Malta, and " in clefts of mountains in Gozo, are heaps " of grey[b] clay, evidently no native of the " soil:" how could the clay have got over the high craggy rocks of those two islands, unless they had been formerly joined to a higher land?

The blocks of granite found at Staffa, at Rugen, and Poel in the Baltic, seem derived from the adjacent land. The Verde di Corsica, the slate, which contains octaedral crystals of iron ore, and other substances, found in the state of pebbles on the coast

[a] The mile-stone, the dykes of Portsea, and Pont-auban.

[b] Boisgelin's History of Malta.

of Liguria, have been referred, with a high degree of probability, to adjacent islands.

In vain did M. de Luc object that these coincidences on the opposite sides of valleys are not found universally; the theory, he rejected, does not require that they should be: it requires only that two places separated by a valley, should be as similar to one another as they might reasonably have been expected to be, had no valley intervened. Where the interval is large, a change may easily take place in the nature, thickness, or position of the strata, and from the unequal hardness and destructibility which we should naturally expect in different rocks, vallies often extend along the lines of junction. In the Orkney Islands may be seen a primitive rock similar to one found at the nearest point of Norway, in contact with a secondary rock similar to one found at the nearest point in Scotland. Had the interval of sea occurred in any other spot, the contact would have been concealed: the opposite coasts would not have been si-

Travels in the North of Europe, p. 10.

milar; but the doctrine of their original union, though unsupported, in that case, by one of the principal arguments by which it is now supported, would obviously have then been neither more nor less true than it is at present.

It is very possible, however, that this argument, founded on the resemblances of distant objects, may be strained by incautious reasoners beyond its just limits. There is a striking resemblance between the simple minerals of Norway and those of the United States of America: but this resemblance, unsupported by other evidence, would hardly entitle us to infer an original union of countries so remote from each other; we are not at liberty to conclude that the Trap of Scotland has ever been connected with that of Bombay, or the coal of Durham with that of New Holland.

At Heidelberg, they say there was once a subterranean communication between a ruined castle which commands the town, and another castle, the site of which is on the summit of a mountain equally elevated on the opposite side of the river. The saying has probably originated in the circumstance of both these castles having a subterraneous passage, now blocked up, in the sand stone on which they are built: as, however, under this sand stone the strata are composed of granite, and the river runs on granite, he who trusts to physical evidence rather than traditional, may safely pronounce this communication fabulous.

Pumice is found on both sides the Rhine, at Andernach: Tarras on both sides the Brohl: Porphyritic lavas on both sides the Gran, in Hungary. Volcanic products occur in many islands in the Mediterranean; yet we are by no means to assume that these are remnants of beds once united.

From the discovery of elephants' bones in the two hemispheres, Buffon hastily inferred that they were once continuous.

It is not only in the nature of the strata that this correspondence in the opposite sides of a valley is observable; it is observable also in their position.

" The vallies of the Jura," says Saussure*,

* Voyages, § 343.

" are often screened by two chains of moun-
" tains, the escarpments of which face one
" another."

At the Dent de Vaulion, the beds dip rapidly to the north; in the opposite valley
to the south. At Besançon, the beds on
each side the valley of the Doux, are so
disposed as if they still wanted to lean
against each other. The strata of the North
and South Downs dip opposite ways. Dorsetshire and the Isle of Wight furnish perhaps more striking examples of similarity
in the posture of strata, their posture being
very uncommon.

Between Fast Castle [a] and Eyemouth in
Berwickshire, may be seen segments of circles in the convoluted strata, the remaining
part of the circles having been apparently
carried away by the cause, whatever it may
have been, which produced the indentations
of the coast.

It is obvious, that correspondence of dip
at two points can afford no presumption
that the rocks which exhibit it have been

[a] Trans. of R. Soc. Edin. vol. vii. p. 81.

united, if in the intervening space we find
a different dip. On the other hand, unconformity of posture does not prove the original absence of continuity, since varieties
of dip occur in uninterrupted strata.

The probability of two opposite cliffs,
coasts, &c. being portions of a rock originally continuous, will be greater or less, as
the points are more or less numerous in
which the resemblance between them can
be established. A correspondence in one
or two respects proves but little ; it may be
the effect of accident ; but a perfectcorres-
pondence throughout renders probability
almost certain. He must be a determined
sceptic who can doubt the pristine connexion of the Isles of Re, Aix [a], and Oleron.
on the western coast of Bretagny, or that
of the several Antilles [b], if, as it is stated, the
same rocks, with the same grain and fossils,
are found in each of them, and the coasts
and hills in all are extended in the same
direction.

[a] l'Encyclopedie Geographie Physique, tom. ii. p.
864.

[b] Ibid. tom. ii. p. 666.

That vallies have been formed by the partial excavation of a mass originally continuous, is further evinced by the fragments generally dispersed over the surface of the earth in the state of bowlder stones, gravel, sand, and other substances, comprehended under the general name of alluvial deposits. These deposits may often be traced to the spot from whence they were originally derived.

At a stream work near Roach, in Cornwall, are found crystals of tin so large, that it became an object to follow them to their source, and the miners have discovered the parent vein.

Huge blocks of Granite are scattered over the plains of Cheshire, Shropshire, Staffordshire, &c. The bed which supplied them may be seen in the Cambrian hills. As this bed, at different places, presents more or less variety in the proportion or aspect of its ingredients, a correspondent variety is observable in the distant insulated blocks, and thus are we enabled to ascertain almost the exact spot from which they have been respectively detached, and the precise route they took in coming to the place which they now occupy.

The Granite blocks on Shap Common may be traced to Westsleddale, those of Kirkby Lonsdale to Burrodale Crag, and those of Kendal to the neighbourhood of High-borough bridge.

With these are found, occasionally, blocks of Greenstone derived from the same district.

In Shalkbeck, Cumberland, the masses which are altogether different from the Granite of the Lakes, may be identified with that of the Criffel Mountain in Dumfries-shire.

In Friesland are low hills, composed of bowlders of Granite, Basalt, Lava, Serpentine, Quarz, and different shisti, accompanying the Dunes which stretch from Zutphen to Arnheim. M. Desmarets [a] has traced these substances to the mountains on the Rhine.

The Granite blocks so extensively dispersed over the North of Germany, have

* Encycl. Geog. Physique, tom. ii. p. 469.

been followed by Von Buch[a], Hausman , &c. to their birth-place, in Norway and Finland.

Those dispersed over Bavaria, Franconia, and Swabia, have been identified by Count Razomowski, with rocks belonging to the mountains of Moravia, Bohemia, and Lower Austria.

As early as the year 1740, Ehrhart[c] had traced to the Tyrol, many of the blocks found in the country situate between the Alps and the Danube. The Granite blocks which lie upon mountains of secondary Limestone, near Gallis, Ostago, Feltrino, Campo de Rovere, and between Astico and the Adige, are likewise recognized by Arduino[d] as belonging to the mountains of the Tyrol.

[a] Geologische Beobachtungen vol. i. p. 19. Travels in Norway, Black's Translation, p. 14. Mem. Berlin Acad. from 1804 to 1811.

[b] Hausman's Nord-Deutsche Beiträge. See l'Encycl. Geog. Physique, vol. i. p. 239.

[c] Philosophical Transactions, 1740.

[d] l'Encycl. Geog. Physique, tom. i. p. 5. and tom. ii. p. 229.

The Granite blocks found near the Lake of Como, are referred by Amoretti[a] to the mountain of St. Gothard.

The huge insulated masses of rock which lie scattered on the mountains of Swisserland, M. Saussure,[b] and, more lately, M. Von Buch[c], have enabled us to follow with confidence to their native beds.

Of the blocks that have lodged on the Jura, the largest, and by far the most numerous, consisting of Granite, were detached, it appears, from the Pic d'Orne, situate at the extremity of Mont Blanc, above the Val de Ferret. Those on the Saleve and Voirons, belonged to the Peaks, which overhang the valley of Montjoie. Others, found near Solothurn, are known, by their peculiar grain, to have come from the Grindelwald.

The Pudding-stone in the walls of Auvernier, Colombier, Corcelles, St. Blaise on the Jorat, and in the Pays de Vaux, is too

[a] Amoretti Viaggio ai tre laghi, p. 175.
[b] Voyages dans les Alpes.
[c] Mem. of Berlin Acad. from 1804 to 1811. See also Journal de Physique, tom. xxx. p. 281.

remarkable not to be recognized immediately as that which occurs in situ in the
Valorsine.

Of the blocks of Gneiss found on the
Jura, some are traced to the foot of Eigers,
others to the mountains between Sem
Branchier and Martigny.

The blocks of black Limestone and
Greywacke are traced to the mountains of
Aigle, the Dent de Midi, and the Dent
de Morcles.

Those of Jade and Smaragdit, occurring near Lausanne, near Maudon, and
the lake of Neuchatel, to the Val de Bagne,
above Sem Branchier ; those of serpentine
in the same neighbourhood to the Glacier
of Durand.

The grey weather stones, so plentifully
scattered over the southern counties of
England, are evidently derived from the
destruction of a rock which once lay over
the chalk. Gravel may, in like manner, be
traced in most instances, to the beds which
supplied it. The flint gravel about London
is derived from two sources at least : it is
in part supplied by the attrition of flints

from the chalk, in part from a regular stratum of gravel interposed between the chalk and the London clay. Mr. Webster supposes, with reason, that this gravel may also contain, occasionally, fragments of upper strata, which formerly existed in this island and have been destroyed. No less true than striking is the remark of the author of a work entitled Contemplations of Nature, that there is no picking up a pebble by the brook-side, but we find all nature in connexion with it.

The resemblances hitherto adduced as occurring on the opposite side of vallies have been brought forward only to prove that those intervals which we now find between mountains, have not been from the beginning ; that vallies owe their origin to the removal of matter which once occupied them ; that there was a time, when, to use a memorable expression of Sir J. Hall, vallies were not only submarine, but subterranean. Another coincidence, still more

remarkable, will not merely corroborate
the arguments already advanced, but evince
further, that the excavating agent was running water.

The disposition of ground often resembles works in fortification, the mountains representing bastions, the vallies covered ways. If we travel along a valley
which runs north and south, the mountains
on our right project eastwards, those on our
left westwards ; and salient and re-entering
angles are to be seen on either side in
alternate order, so that the one shall be
invariably opposed to the other.
After having crossed the Alps thirty
times, the Appenines twice, and having, by
repeated excursions familiarized himself
with the entire chain of the Jura, Bourguet
published this doctrine in 1729, and expressed surprize that a phenomenon so
obvious and striking should have remained
till then unobserved. Fully aware of the
importance of his discovery, M. Bourguet
entitles it a Key to the Theory of the Earth.
His doctrine does not appear, however,

to have received from the public the attention it deserved, till illustrated and enforced by the eloquence of Buffon.

The windings in the channels of rivers, says this celebrated naturalist, " have cor- " responding angles on their opposite banks, " and as mountains and hills which may be " regarded as the banks of valleys, have " likewise sinuosities with corresponding " angles, this circumstance seems to demon- " strate, that vallies have been formed in " the same manner as the channels of ri- " vers."

Is this rule capable of being generalized? The author whom I have just quoted, observed, that the correspondence held equally good, whether the mountains were separated by an extensive plain, or narrow valley. He found corresponding angles, like those of inland mountains, on the opposite coasts of the straights of Magellan; and Tournefort, on the opposite sides of the Hellespont; but Humboldt[a] has given to this doctrine much greater extension: He says " the salient and re-entering

[a] Journal de Physique, tom. liii. p 32.

" angles of Europe, Africa, and America,
" evince, that the old and new Continents
" have been divided by the action of water,
" and that the Atlantic is a valley so exca-
" vated."

In looking at a map of the world, it is
not difficult certainly to discover occasional
appearances of corresponding bays and
promontories. The coast of Guinea forms,
if you please, a salient, and that of Congo,
a re-entering angle, commensurate with the
re-entering angle of Mexico, and the salient angle of Brazil; but does the correspondence extend along the whole line
of continents, or is it of such rare occurrence, that the rule is lost in the exceptions? If this reciprocity of form,
did, indeed, pervade the coasts of opposite
hemispheres, we should have reason to
doubt, whether it might not have arisen in
other places, from a different cause than is
now assigned to it. Upon what do the
sinuosities of running water depend? upon
its oscillation; upon its falling away alternately from one side to the other, in consequence of obstacles; upon its tendency to

move from parts that resist more, to parts
that resist less. We are not, therefore, to
expect traces of such a motion on the opposite coasts of the Old World and the
New, unless we suppose that a body of
water, having power to scoop out the bed
of the Atlantic, has met with some impe-
diment sufficient to divert its stream from
the coasts of Europe and Asia, to that of
America. Accordingly, so far from being
able to discover alternate and opposite
angles along the shores of the Ocean, we
cannot discover them along the shores of
the Mediterranean, the Baltic, the Red
Sea, the English or Bristol Channel, nor
even along the banks of the inland lakes of
Windermere, Loch Lomond, or Killarney.

Other writers deny the fact to be so general, even as Bourguet represented it to
be ; in the vallies of the infant Rhine,
Rhone, and Reuss, they have searched for
these appearances in vain, and, therefore,
suppose them confined to secondary districts : but it is not true that they belong
more to secondary districts than primary,
nor if true, would the instances adduced

affect the general reasoning. Mountain torrents scarcely oscillate ; they either tear and carry off the obstacles they encounter, or shoot headlong over them. It is not, therefore, amid the rapids of Alpine countries that we should expect sinuosity, if the theory of aqueous excavation were true ; and its truth is the more probable, from our not finding sinuosity in such situations. The straightness of the channel of a river depends in great measure on the rapidity of its stream, the curves being few and sharp, where the declivity is steep ; numerous, easy, and swelling, where it is gentle. If, in primitive districts situate on low ground, traces of sinuosity are not discernible, while they are distinct in secondary rocks situate at a higher level, what other inference can be drawn, than that some substances have yielded from their softness to those impressions, which others by their hardness have been enabled to resist ?

The savages [a] in North America are said to be so far advanced in natural science as

[a] Buffon, Smellie's Translation, vol. i. p. 257.

to form pretty accurate computations of
their distance from the sea, by observing
the courses of rivers : if a river runs nearly straight, for fifteen or twenty leagues,
they know themselves to be a great way
from the coast : if there are many sinuosities, they conclude that they are not far
from the sea.

Valleys have a tendency to increase gradually in breadth as they descend, modified
however by two disturbing causes, which I
shall presently have occasion to notice. A
necessary consequence of this tendency is
the gradual tapering of promontories. The
wedgelike form, so strikingly exemplified in
the peninsulas of Africa and South America,
is found for the most part in individual hills
and mountains, the broad end fronting the
head [a] of the valley.

Now in regard to disturbing causes :—

The breadth of valleys depends in
some measure on the comparative hardness
of the substances which bound them. The

[a] Transactions R. S. Edin. vol. vii. p. 170.

great vallies of Hungary and Bohemia consist chiefly of soft secondary rocks ; but at
the lower extremity these vallies contract
into gorges, because in that part of their
course they are bounded by primary rocks.

Again, the size and direction of a valley
change as often as it is joined by lateral
vallies, and the amount of the change so
produced varies in proportion to the size of
the valley that produces it, and the angle at
which the two meet.

The larger the valley, the more even is
in general the surface of its bottom, this
evenness resulting from the accumulation
of debris.

———

From all these considerations, I think we
are justified in concluding that vallies have
in general been formed by the action of
running water ; and consequently, that
mountains in general are not the effect of
volcanoes *, as Lazzaro Moro, Stenon,

* For a more detailed refutation of these opinions
vide Deluc's Lettres Phys. Disc. 47. — Catcot on the
Deluge, p. 178. — Bertrand Nouveaux Principes de
Geologie p. 3 and 35, and Cuvier's Discours Preliminaire.

Sprengseysen, and Kruger supposed; nor
of earthquakes, according to the doctrine of
Ray and Hook; nor accumulations of sand
or mud brought together by submarine
currents, as was imagined by Le Cat, Buffon,
Le Maillet, and the Bishop of Clogher;
nor crystalline shoots, as Rouille and La
Metherie represent them; nor remnants of
subsided strata after the notion of Deluc
and Hollman[a]; but the hardest and least
destructible portions of the earth, as it stood
at some earlier epoch.

The terms mountain and valley are relative; that, which is mountain compared
with the ground beneath, is valley compared with the ground above it. The valley of
the Thames at London is contained in that
of which Clapham Rise forms part of the
boundary on one side, and the Green Park
on the other, and this again is contained in
the larger valley which occupies the interval between Highgate and Sydenham.
Arrived at these points we find our horizon

[a] Journal de Physique, Introd. tom. ii.

bounded by a chalk ridge still loftier. In like manner, continents are made up of successive steppes, or terraces, rising on every side from the sea to the summits of Mexico and Thibet, so that, with the exception of these, every spot upon the globe is overlooked by some other : but whether we consider the large or the small, the including valley or the included, the first member of the series or the last, the phenomena are uniform in kind, the course of the mountains being always determined by that of the valleys, the course of the valleys by that of the excavating waters.

———

The truth of these opinions will appear still more evident if we consider the phenomena of what are called alluvial deposits.

1. These deposits, whether found on hills or in valleys, seem to have been invariably derived from the breaking up of rocks, situate at a higher level than themselves.

2. The larger masses of the same substance, are generally found nearest to the parent rock.

Dr. Ehrhart[a] speaking of gravel in the Tyrol says, " these stones increase " in bulk from Memingen towards the " Alps, till they get to be three or four " feet in diameter ; in the opposite direc- " tion they gradually decrease to the size " of coarse sand. We may collect from " Guettard[b] that a similar gradation is " found in the gravel which covers the " plains of Poland, from the Carpathian " mountains to the Baltic."

True it is, that blocks of very different sizes are sometimes found together, both on hills and plains ; that in some places the small pieces are in abundance, though there are few blocks, and that in others the blocks are in abundance, though there are few of the smaller pieces.

But these seeming irregularities, referable to some local cause, only confirm the

[a] Phil. Trans. for 1740.
[b] Mem. Acad. des Sciences for 1762. Playfair's Illust. p. 382.

argument which they have been adduced to disprove [a]; for similar irregularities take place on a smaller scale, in the detritus produced on the sea-shore by the waves, and by torrents inland.

3. Those blocks or pebbles which are most distant from their native place, are composed of the hardest and most indestructible materials as Granit, Greenstone, and Quarz, Chert, Flint, and Jasper. It is in part owing to their rapid disintegration, that Basalt does not yield gravel in the vallies of Antrim, nor Chalk in the Weald of Kent and Sussex, nor Oolite more abundantly in the vales of the midland counties.

4. Substances which break into cubic or hexagonal blocks, are found at a greater distance from their native place than those which break into blocks, the angles of which are acute; this is one reason why Granit bowlders have travelled further than slate.

[a] Deluc's Travels, vol. i. p. 121.

The occurrence of these bowlder stones in Switzerland, and along the shores of the Baltic, is notorious; but the phenomenon is by no means confined to these countries. At Glenmalur, in the county of Wicklow, huge masses of Granit rest upon the mica slate. Along the vallies of the Garonne and Gave de Pau, you find granitic blocks derived from the Pyrenees; along the valleys of the Aveyron [a] and Dordogne, from the Cevennes; at Bains from the Vosges. In the department of Morbihan[b], the number of blocks is estimated at four thousand, and some of them are not less than twenty feet in height.

Near Turin[c], the calcareous hills are covered by blocks of Granit, some of them of the size of thirty cubic feet, although no mountain of that substance is found within many leagues. Gerenna, in Grenada[d], is famous for its bowlders, which suggest the idea of a shower of stones.

---

[a] Bertrand. Nouveaux Principes de Geologie, p. 161.
[b] Cambry, Monumens Celtiques.
[c] La Metherie, Theorie de la Terre, tom. iv. p. 417.
[d] Bowles.

The mountain of Oden Tschelonn[a], in Siberia, according to Patrin, owes its name (which in the Mogul language signifies petrified flocks) to the Granit bowlders which are found there.

La Metherie[b] says, there are none of these blocks in Asia; it is more probable that, insensible of their importance, travellers in that part of the world have passed them by unnoticed.

Near the Lake Asphaltites[c], are blocks which have been mistaken for mutilated statues.

Chardin found in the plains of Media, stones so large, that it would require at least eight men to move any one of them, and yet there is no stone, he says, of the same kind in *situ*, within a circuit of eighteen leagues. In the mountains of Arabia, near Angoura, mention is made of small pyramids, which I suppose to be bowlders. Paul Lucas [d] states their number at twenty thousand.

[a] Journal de Physique, tom. xxxix. p. 339.
[b] Theorie de la Terre, tom. iv. p. 419.
[c] Volney's Travels in Egypt.
[d] Paul Lucas, tom. i. p. 160.

Of the famous rock in Horeb, said to be that which at the touch of Moses' rod furnished water to the Israelites in the wilderness, Dr. Shaw [a] gives us the following account : " It is a block of granite marble " about six yards square, lying tottering as " it were in the middle of the valley, and " seems to have belonged to Mount Sinai, " which hangs in a variety of precipices " all over the plain." I am informed that blocks of Granit extend for more than one hundred miles on the south of Lake Huron, in North America, and appear in islands twelve miles from its margin.

Granit bowlders, therefore, are not of partial occurrence, nor is the theory tenable, which supposes those found in the North of Germany, to have slid thither upon the ice.

A late naturalist [b], who, dying in the fullness of years, left behind him a name much too respectable to prevent his errors from being contagious, advanced a very extraordinary hypothesis, to explain the

[a] Shaw's Travels, p. 352.
[b] Deluc's Geol. Tr. vol. i.

blocks so frequent on the Jura, and in Northern Germany ; he supposed these blocks to have been thrown up by the expansive power of Gas, generated at the time of their formation, and to have fallen where we now find them ; that is, resting upon beds of limestone and sandstone, the pedestal on which they rest unshattered. How blocks of such enormous weight and magnitude, could fall upon beds so fragile, without fracturing them, it is not easy to discover ; still less, how such an event could happen before these beds were in existence ; for, I suppose, no one will claim for the mountains of Jura so high an antiquity as is conceded to Mont Blanc.

It is some palliation, however, of this hypothesis, that it was constructed at a time when the imaginations of all men were so dazzled by the brilliant discoveries then making, in pneumatic chemistry, that it was almost as difficult to speculate without Gas, as to breathe without air.

The circumstance of primitive blocks resting so frequently upon secondary beds, furnishes an argument equally conclusive against the opinion that these blocks are only the survivors[a] of a catastrophe by which the adjoining parts of the strata to which they belonged were destroyed.

These theories refuted, there remains, in explanation of the phenomena of bowlder-stones, the theory which attributes their occurrence, like that of ordinary gravel, to the action of running water.

The arguments in favour of that doctrine are, that bowlder-stones are evidently not *in situ* ; that they are, for the most part, traceable to the parent rock, which, however distant, is always at a higher level than themselves; that they often rest upon beds either secondary or alluvial ; and lastly, that the upper surface of rocks protected by soil, is in many cases so furrowed[b] as to resemble a wet road, along which a number of heavy and irregular bodies have been dragged, these furrows ge-

[a] Bertrand's Geologie, ed. i. p. 160. et seq.
[b] Vide Transactions R. S. Edin. vol. vii. p. 139. where this circumstance is described at length and exemplified.

nerally agreeing, in parallelism, both with one another, and with the ridges and large features of the district.

On the other side, it is objected, first, that between [a] the supposed parent rock and the bowlder-stones, there is often an interval in which none of these bodies are seen ; secondly, that vallies, rivers, lakes, arms of the sea, intervene between the bowlder-stones and rock with which they are supposed to have been originally connected ; thirdly, that these stones are often much too large to have been swept along by the action of water.

The first of these objections, however, proceeds on an assumption which I apprehend to be altogether gratuitous, since it has not been shown, that, had bowlder-stones been so transported, there would have been no vacant intervals, or that their distribution would have been in any respect different from what it actually is.

The second objection does not apply to bowlder-stones having been transported by the action of running water, but simply to

[a] Deluc's Geological Travels, vol. i.

their having been so transported at a period subsequent to the formation of vallies, rivers, lakes, and arms of the sea.

To estimate the value of the third objection, it is necessary to consider separately what is the magnitude of the largest of these blocks, and what power running water possesses, of removing them from one spot to another.

The dimensions of the block, out of which was hewn the pedestal of the statue of Peter the Great, after being somewhat reduced, were, length at the base 42 feet; at the top 36 feet; breadth 21 feet; height 17 feet: Its weight exceeded 1500 tons.

The Needle[a] Mountain, in Dauphiné, said to be a Bowlder, is one thousand paces in circumference at the bottom, and two thousand at the top. At Pierre à Bot[b], above the lake of Neufchatel, is a Granite block forty feet high, fifty long, twenty broad, weighing thirty-eight thousand cwt. The block called Pierre à Martin[c], measures

[a] Hist. de l'Academie des Sciences, for 1700, p. 4.
[b] Memoirs of the Berlin Academy, from 1804 to 1811.
[c] Transactions of the R. S. Edin. vol. vii. p. 142.

ten thousand two hundred and ninety-six cubic feet.

In what manner can running waters have acted, so as to set in motion and transport to very considerable distances masses of such prodigious dimensions?

The author [a], by whom the theory of Dr. Hutton has been so ably illustrated and defended, always disposed to cut the knot, rather than call upon the gods to untie it, has referred this extraordinary phenomenon to the operation of ordinary causes. After noticing the celebrated blocks scattered upon the surface of the narrow vale or glen which separates the Great from the Little Saleve, (one of them measuring about 12 hundred cubic feet,) " for my part," he says, " I have " no doubt that the Arve, which is still at " no great distance, when it ran on a higher " level, and in a line different from the pre- " sent, aided by the glaciers and superior ele- " vation of the mountains, was an engine suf- " ficiently powerful for effecting the trans- " portation of these stones."

Another eminent disciple [b] of the same

Illustrations, p. 392.

Transactions R. S. Edin. vol. vii. p. 142.

school ventures upon this point to dissent
from the opinions of his master. " To move
" a mass of granite of even fifty or sixty
" cubic feet, and to carry it a few yards,
" would," he says, " require the utmost ef-
" forts of the Rhone or the Arve, as they
" flow near Geneva, in their highest floods ;
" but that such blocks could be conveyed
" by one of them along its whole course, is
" contrary to all experience, and still more
" when we consider that these rivers are di-
" vided, at their source from beneath the
" Glaciers, into forty or fifty small streams.
" Yet from the Glaciers these blocks must
" have come ; and when we take into ac-
" count the magnitude of some of these gra-
" nitic masses, it is clear that the task is
" beyond the power of any River that flows
" on the surface of the earth."

Bowles, the traveller in Spain, who bestowed much attention on this subject,
thinks, that Rivers, flowing under ordinary
circumstances, are incompetent to transport to any distance, not only colossal
blocks, but moderately-sized gravel.

" From the singularity of their appear-

K 3

" ance," he says, " there are few pebbles
" which it would be so easy to recognize,
" as those in the bed of the Henares, near
" St. Fernandez. If they ever moved at all,
" they ought, in the course of ages, to have
" found their way into the Tagus a little
" way off ; but there is not one of them in
" the Tagus.

" At Sacedon, the Tagus is full of lime-
" stone pebbles : lower down, at Aranjuez,
" are none. Nobody has ever seen granite
" pebbles, large or small, in the Ebro, nor
" blue stones veined with white ; yet the
" Cinca, before it joins the Ebro, abounds
" in them.

" White and red pebbles of Quarz are
" found in the bed of the Noxera, which
" likewise falls into the Ebro ; but in the
" Ebro is found nothing of the kind. The
" Guadiana in different parts of its course
" flows over pebbles, similar to those
" found in the strata of the adjacent hills ;
" but those, which occur half a league up
" the stream, never mix with those which
" occur half a league down ; and at Bada-
" jos, stones of this kind, being no longer

" found in the cliffs, are no longer found
" in the river.

" At the source of the Loire are pebbles
" innumerable; lower down, at Nevers,
" only sand.

" In the Yonne river, above Sens, are
" flints in abundance; for they abound in
" the banks of the Yonne, about Joigny.
" The Yonne falls into the Seine above
" Paris; but who ever saw any of these
" flints at the Pont-neuf, or any pebble
" whatever, round or angular?

" Near the Perte du Rhone you cross
" the river of the Valorsine, which is full
" of pebbles, because the country it flows
" through is full of them. At one place,
" this river tumbles into a kind of cavern;
" If pebbles were carried down by Rivers,
" the cavern ought to contain them in
" abundance; It does not contain one. On
" my way to Geneva, I threw some stones,
" which I had marked so that I might
" know them again, into this river, just
" above its fall; and there I found them on
" my return; They had not advanced an
" inch during my absence.

K 4

" The Rhone, Garonne, and Adour rivers,
" remarkable for the quantity of pebbles
" they run over in one part of their course,
" have only sand at their mouth."
These remarks, being equally applicable
to all countries, may be verified by every
one in his own neighbourhood. I therefore
abstain from adducing the similar instances
which crowd upon my mind, when I consider what happens in our English rivers.
Enough has been said to prove, that, flowing, under ordinary circumstances, over
ground which is level, or nearly level, the
power, which Rivers possess to propel even
gravel, is so trifling as scarcely to deserve
consideration. When, their stream being
accelerated by the near approach of their
banks, by the occasionally increasing declivity of their beds, or by the augmented
volume of water afforded by heavy rains or
thaws, they exhibit themselves in the form
of a Torrent, it cannot be denied that they
have the power of carrying, to a certain
distance, the fragments which havè fallen
from adjacent cliffs or hills, and of fretting
and rounding them by mutual attrition ;

but even under these circumstances, I believe, he, who can divert his mind from the picturesque beauty of a cataract to a philosophical consideration of its effects, will be disposed to admit that the power of Rivers, when most impetuous, is considerably less than is in general apprehended. At all events it is but short-lived, and ceases with the cause.

If then we chuse to suppose that the blocks scattered so extensively over the face of the Jura chain were brought by Rivers, they must have been brought by rivers, descending with great velocity along a plane, regularly inclined to the distance of many leagues; but, supposing the inclination to be the smallest which the advocates of this theory consider capable of effecting the object, it has been clearly shown by Sir James Hall, that the summit of such an imaginary plane would be far above the level of perpetual snow, and, consequently, far above the level at which rivers can exist.

The phenomena of the blocks, which have been traced to the northern parts of Europe, are still more adverse to the hypo-

thesis of Dr. Hutton. The blocks derived
from the Alps are not known to have travelled, in any instance, more than sixty
miles, whereas, those which came from the
mountains of Scandinavia, are said to have
been traced to at least seven times that distance. Add to this their lateral, which is
proportionate to their longitudinal extent ;
They are scattered over the Continent from
Holland to Petersburg and Moscow. It is
perfectly incredible, that these blocks, extending over so immense an area, and found
on the opposite sides of lakes and seas,
should be the waifs and estrays, either of a
single River or of any number of Rivers.

———

If the Transportation of bowlder-stones
cannot be referred to the agency of Rivers, so neither can the Excavation of
Valleys.

1. Some valleys are dry, as that of Ched-
dar; the valley of Rocks near Linton; the
Winyats near Castleton in Derbyshire.
How can it be supposed that these have

been excavated by a River which has no existence ?

2. If Valleys were formed by the Rivers that flow in them, how happens it, in so many instances, that the source of a river should be below the head of a valley?

3. In very many cases, where pits have been sunk to considerable depths on the banks of large rivers, the alluvial land has reached far below the level of their bed. This circumstance, to which we are to ascribe the flatness of the valleys through which they run, is altogether inconsistent, as Mr. Playfair acknowledges, with the notion, that in these places the bed of the river was excavated by the River. The action of Rivers may consist, M. Cordier observes, either in filling up or in scooping out; it cannot consist in both; if in scooping out, it has not formed the Beds of Gravel; if in filling up, it has not formed the Valleys.

4. It has been already remarked, that every valley is included in a valley still

larger. Grant that the interior has been formed by a River, can the exterior ones be ascribed to the same cause ?

5. The circumstance of Rivers changing their bed, shows how little they are adapted for the purpose of excavation : if they formed a bed for themselves they could not change it.

6. To suppose that Rivers formed their own banks, is to suppose that rivers were once without banks, a supposition evidently absurd.

Buffon conceived that the inequalities on the surface of the earth had been occasioned, principally, by the motion of a Sea that once covered it ; but the action of a sea on its own bed appears altogether incapable of producing such an effect ; it consists in little more than shifting sand and other unconsolidated matter from those places, in which the current is strong, to those in which it is feeble or null.

At great depths, where there is no motion there can be no abrasion. Lamanon came much nearer the truth when he suggested, that submarine valleys existed before the existence of the ocean, and that currents were rather the effect of these valleys than the cause of them.

But is the Sea really incompetent to produce these effects? The striking resemblance between inland cliffs and those upon the coast affords strong presumptive evidence that both are owing to the same cause ; and surely cliffs upon the coast are produced by the assaults of the Sea. He, who travels along the shores of this country, will find numerous instances of towns and churches, situate too near this destructive agent, having fallen a prey to its devastations, and many others, once inland, now no longer so, already undermined, maintaining only a precarious [a] existence. The concurrent testimony of nations assures us of ravages by the Sea, far more extensive than any of which we can obtain a knowledge by means of personal observation.

[a] See Philosophical Transactions for 1716. No. 349.

The Antiquarians of Cornwall undertake to prove from the authority of old charters and parchments, that a tract of land once extended from St. Michael's Mount to the south of Penzance; a considerable town, they tell us, standing half way between the Land's-end and the Islands of Scilly[a], has, with the circumjacent country, been overwhelmed by the waves. The Causeway of St. Patrick is pointed out to us, in another part of the coast, as the only memorial, which nature has left, of a tract of land, which connected England with her sister kingdom, till the connexion was broken off by the Sea. Conflicting tides are said to have rent England from France[b], Denmark from Sweden. It is owing to the encroachments of the Atlantic, if Strabo and Pliny are to be believed, that Spain[c] is separated from Algiers; Sicily from Apulia and Tunis; Corsica from Piedmont and Sardinia; Italy

[a] Borlase.

[b] Wallis. Buffon, Smellie's Translation, vol. i. p. 489 and 496.

[c] Pliny, l. iii. c. 8. Sicilia quondam Brutio agro cohærens, mox interfuso mari avulsa.

from Greece; Crete from the Morea; Constantinople from Asia Minor. We are told that the Chinese Islands[a], the Philippines, Borneo, Java, New Guinea, New Holland, are only Islands in consequence of the force successfully exerted by the Gulph Stream upon the once intervening country. The Maldives[b], Ceylon, and even Madagascar, are said to be mere remnants of a territory which extended formerly from the promontory of Africa to that of Hindostan; and the Atlantis of Plato is supposed by his commentators to refer to the submersion of another territory, no less extensive, situated where the waves of the Atlantic now roll between America and Europe.

Leucada continuam veteres habuere coloni,
Nunc freta circumeunt. Zancle quoque juncta fuisse,
Dicitur Italiæ, donec confinia pontus
Abstulit, et mediâ tellurem repulit undâ.

OVID.

Not one of these traditions, however, is entitled to the smallest credit; they are

[a] Buffon, Smellie's Translation, vol. i. p. 177.
[b] l'Encyclopedie, Geographie Physique.

unsupported by evidence, opposed to all
our experience, and must be accounted the
mere arbitrary speculations of men, who,
anxious to explain the correspondence,
which they had noticed, between distant
shores, rashly ascribed to an agent, of which
they did not know the power, events of
which they did not know the existence.

Dr. Hutton caught the contagion of ancient and popular error; he did not fail,
however, to perceive, that the depreda-
tions committed by the sea, had been
greatly over-rated. " The description,"
he says [a], " which Polybius has given
" of the Euxine, with the two oppo-
" site Bosphori, the Mæotis, the Propon-
" tis, and the port of Byzantium, are as
" applicable to the state of things now,
" as they were at the writing of that his-
" tory. The Isthmus of Corinth is appa-
" rently the same at present as it was two
" or three thousand years ago. Scylla and
" Charybdis are still, as in ancient times,
" rocks hazardous for coasting vessels; the
" Port of Syracuse, with the Island which

[a] Theory of the Earth, vol. i. p. 190.

" forms the Greater and Lesser, and the
" Fountain of Arethusa, the water of which
" the ancients divided from the sea by a
" wall, do not seem to be altered. From
" Sicily to the coast of Egypt, there is an
" uninterrupted course of sea for one thou-
" sand miles ; consequently the wind
" should bring powerful waves against
" those coasts. But on this coast of Egypt,
" we find the rock on which was built the
" famous Tower[a] of Pharos, and at the
" eastern extremity of the port Eunoste,
" the Sea Bath cut in the solid rock upon
" the shore. Both these Rocks, buffeted
" immediately by the waves of the Medi-
" terranean sea, are to all appearance the
" same at this day, as they were in ancient
" times."

" Shoals, the terror of seamen," says
Dolomieu, " do not perish by time ; co-
" vered perpetually by the foam of the

[a] Savary, Lettres sur l'Egypte. How slight are the
changes which have taken place on the surface of our
planet within the limits of historical record may be seen
in l'Encycl. Geog. Phys. vol. ii. p. 561. Guettard's Memoirs, vol. iii. p. 209 & 223, and Cuvier's Discours Preliminaire.

L

" waves, they still maintain the conflict;
" and, after the lapse of one thousand years,
" a new shipwreck, in the same spot, attests
" how little change has taken place within
" that period. Not an instance can be
" produced of a rock, a mile in extent,
" having been washed away, during all the
" ages which have elapsed since history
" began."

If the Sea has been supposed to overflow entire continents, it has been supposed also to abandon them.[a] An old saying of Pythias has been quoted, to prove that Sweden was once made up, like Denmark, of Islands; and the Caspian Sea is stated, on the authority of Strabo and Pliny, to have communicated formerly with the Baltic. These suppositions, however, like their opposites, are unworthy of credit; the Sea is no more capable of such extraordinary retreats, than of such extraordinary inroads. In violent storms, earthquakes, &c. it may

---

[a] On the retreat of the sea vid. l'Encycl. Geog. Phys. tom. iv. under the titles Meander and Palus Mæotis. Playfair's Illustrations, p. 441. Buffon, vol. i. p. 492. of Smellie's Translation.

produce consequences highly important to the property of individuals, or even the happiness of districts, but which are of no account as affecting the general structure of the globe; in its diurnal movements it fills up [a] bays with sand, and undermines promontories for a time, till at last the fallen rubbish, forming a beach, guarantees them from further destruction. It is obvious then, that agents, so circumscribed in their operations as Seas and Rivers [b], are little calculated either to effect the transportation of the bowlder-stones so often mentioned, or to produce those inequalities of mountain and valley, which the surface of our earth presents.

But, replies Mr. Playfair, " if the causes " assumed appear inadequate to the effects " produced, it is only because, in respect to " man, their movements are too slow to be " perceived. The utmost portion of the " progress to which human experience can

---

[a] I believe M. Deluc is the first author who entertained correct opinions on these subjects.

[b] M. La Marck, in his Hydrogeognosie, attributes to these causes the formation of mountains.

" extend, is evanescent in comparison with
" the whole, and must be regarded as the
" momentary increment of a vast progres-
" sion, circumscribed by no other limits
" than the duration of the world. Time
" performs the office of integrating the in-
" finite small parts of which the progres-
" sion is made up, it collects them into one
" scene, and produces from them an amount
" greater than any that can be assigned."

Ye Gods! annihilate, but Space and
Time! was the pious but foolish and happi-
ly ineffectual exclamation of a lover who
thought, that, under such circumstances, he
should be happy. Ye Gods! perpetuate
Time! says the Plutonist, and thinks his
reasoning will be incontrovertible.

But suppose the prayer granted; suppose
the Plutonist to have at command whatever
time he desires; Time graduating into eternity; nay Eternity itself; what use could he
make of it? what profit can a man expect
from putting Zeros out to interest? what
increase of weight from a Fast sufficiently
prolonged?

If Seas and Rivers do not tend to produce,

within the period of human experience, any such effect as that which we are endeavouring to account for, they will evidently produce no such effect in a million of centuries. Time may complete that which is in progress; it will never complete that which can never be begun.

The Plutonists should therefore be required to make out a stronger case than they have done, before they are allowed that exemption from the Statutes of Limitation which has hitherto been granted only to the king and the church.

If Seas and Rivers are, from their feebleness, inadequate to produce the effects which have been produced by the action of water, the only remaining cause, to which these effects can be ascribed, is a Debacle or Deluge.

Of those by whom this principle is admitted, some, as Pallas, Lamanon, Sir James Hall, suppose the Debacle to have been partial; others, universal.

That the partial Inundations to which every country is more or less exposed from

Earthquakes, Water-spouts, the Melting of Snows or Glaciers, and the Interruptions produced by the Fall of Mountains, or any other cause, which rivers occasionally experience in their progress towards the sea, are inadequate to the explanation of such phenomena as we have been describing, is too obvious to be insisted upon, and, that the same objection applies to the Deluges imagined by Lamanon and Sulzer, as derived from the Overflowing of Lakes, the reader will immediately perceive, who shall propose to himself the following questions :

1. What examples have we of Lakes overflowing ?

2. How could the overflowing of Lakes produce those great valleys in which the lakes themselves were situate ?

3. How could Lakes exist unless there were already higher ground to embank them ?

4. How could Lakes escape unless there were already lower ground to overflow ?

5. Valleys take their rise on every side of mountains. Did the Lakes supposed

to have formed these valleys burst on every side ?

6. What must have been the extent of that Lake, the bursting of which hollowed out the Atlantic ? or rather formed a valley from Caucasus to the Andes, and from the Andes to Caucasus ?

7. What must have been the power of that Lake, the bursting of which transported the granite blocks of Mont Blanc to the Jura ? of Finland to Silesia ?

8. What reason have we to ascribe to Lakes an origin anterior to that of seas and rivers ?

Of the partial Deluges invented by Pallas and Sir James Hall, some, at least, may be considered almost universal ; for, being derived from the sea, they are supposed to have over-topt the Alps [a], and the mountains of Tartary : One such Deluge would perhaps explain all the phenomena which

[a] It is possible that I here misrepresent, though unintentionally, the opinion of Sir James Hall, and that he considers the transportation of the Jura blocks to have taken place under the sea.

we want to explain; and if these authors have admitted a plurality of deluges in defiance of the recommendation of Newton, not to multiply causes unnecessarily, the reason is, that, in order to obtain their supply of water, they have had recourse to agents which left them no other choice than that of doing their work over and over again, or not doing it at all.

The cause, to which Pallas ascribes his Debacle, was the Shock experienced by the sea during those tremendous eruptions which gave rise to the Moluccas, Philippines, and other Volcanic Islands situated in the Indian Archipelago. His object in imagining this Debacle was not to explain the inequalities met with on the surface of the earth; the symmetrical construction of mountain and valley; the phenomena of bowlder-stones; the distribution of alluvial deposits equally over islands and continents; but to acount for the discovery of the Bones and Tusks of Animals inhabiting the southern latitudes in the frozen regions of Siberia. Now, 4

1. It is scarcely conceivable that these Bones should have travelled from the Indian [a] to the frozen ocean, a distance of 36,000 miles, without fracture or abrasion.

2. Not only insulated Bones of Elephants have been found in those northern latitudes, but their Skeletons, their very Skins and Hair. Had these animals been transported so far, though moving at the rate of 100 miles per day, still they would have been in a state of putridity long before their arrival at the places in which they are actually found.

3. With the bones of the Elephant [b] and Rhinoceros are intermixed those of the Elk, Gazelle, Horse, Ox, Buffalo, animals which inhabit northern climates.

4. Granting such eruptions to have taken place, there seems no reason why the current occasioned by them should have

[a] Journal de Physique, tom. lix. p. 244.
[b] Journal de Physique, tom. lxxx. p. 46.

taken a northern rather than a southern direction.

5. The rising of these islands could displace only a quantity of water equivalent to their bulk, a quantity altogether inadequate to the task assigned to it, that of surmounting the highest chains of Asia.

Sir James Hall was induced to adopt the theory of a Debacle, principally by a desire to account for the granite blocks dispersed over the Jura : but the cause to which he supposes that Debacle to have been owing, is stated too generally to admit of a detailed refutation. In vain does he tell us that granite is of a more recent date than the rocks with which it is asssociated ; that it has been thrown up by Plutonic explosions; that continents have been elevated by similar explosions ; unless he tells us also, what continent was raised at the time the Debacle took place, and where the granite is to be found the forcible ejection of which occasioned the elevation of that continent.

The universal diffusion of alluvial sand, gravel, &c. proves that, at some time or other. *an* Inundation has taken place in all countries; and the presence of similar alluvial deposits, both organic [a] and inorganic, in neighbouring or distant Islands, though consisting often of substances foreign to the rocks of which the islands are respectively composed, makes it highly probable, at least, that these deposits are products of *the same* Inundation.

The universal occurrence of mountains and valleys, and the symmetry which pervades their several branches and inosculations, are further proofs, not only that *a* Deluge has swept over every part of the globe, but probably *the same* Deluge.

The next argument, which I shall advance in support of this conclusion, is founded on an almost invariable want of correspondence between the figure of the surface and the disposition of the strata or veins beneath it. Though where faults occur, the strata are tossed and turned in all directions, " it is " extremely rare," says Mr. Farey, to

[a] Playfair's Illustrations, p. 461. La Metherie's Theorie de la Terre, tom. v. p.197.

*l 6

" find a lifted edge or corner of strata,
" standing up above the general surface;
" the faults, however large the rise which
" they occasion, being rarely discernible by
" any sudden inequality of the ground : nu-
" merous as cliffs, facades, mural ascents or
" precipices are, very few of them are ow-
" ing to faults ; in general, the matter has
" been carried off." [a]

[a] Mr. Hutchinson, who wrote about the middle of the last century, and of whose geological opinions I have more than once had occasion to speak with much respect, was the first author, I believe, by whom this important fact was noticed. Works, vol. xii. p. 338. It has since appeared in the works of Catcott (on the Deluge, p. 165,) Williams (Mineral Kingdom, 2d edit. vol. i. p. 338,) and Desmaret (Geographie Physique, vol. ii. p. 551;) it is noticed also by Mr. Playfair and M. de Luc, and is much insisted upon by Dr. Richardson and the author of the Survey of Derbyshire, from which the above passage has with some abridgment been extracted.

Buffon was sadly mistaken on this subject when he said " I have often remarked, that when the top of a " mountain is level, its strata are likewise level ; but when " the top is not horizontal, the strata follow the direction " of its declivity. It has frequently been alleged that " the beds of quarries incline to the east : but in all the " chains of rocks which I have examined, I found that " these beds always follow the declivity of the hill, whe- " ther its direction be east, west, south, or north." (Smellie's Translation, vol. i. p. 172.)

Cor. 1. Hence the conformity between the direction of mountain chains and that of the strata composing them, is not, as Humboldt supposes, necessary — but only accidental.

Cor. 2. Mountains are not owing, as Deluc thought, to a subsidence of the strata which occupied their intervals.

———

A general view of the structure of our globe, if taken with accuracy, would tend perhaps still farther to convince us of the universal operation of this Deluge.

The southern coasts of the German ocean and Baltic, together with the north of Asia, consist of Marshes, Sands, and Alluvial plains, the rivers of which flow indiscriminately in almost every direction. In this enormous tract of *low or smooth* land, rise the Scandinavian, Uralian, English, Welsh, Scotch, and Irish Mountains, forming insulated groups: Such, under the same latitude, seems to be the structure of North America.

The central part of the old Continent exhibits a girdle of *rough or Alpine* land, extending from Portugal to China. Prodigious Mountains, many and deep Lakes characterize this tract, over a large portion of which Volcanos are distributed.

In the opposite part of the new Continent, we find the highest Mountains of America, the Gulph of Mexico, and the Volcanos of the West Indies.

South of this Alpine country, a range of sandy Desert extends over the old Continent with few interruptions, from the shores of the Atlantic Ocean to those of the Pacific. Opposite to this, on the map of America, we find the vast Plain of the Amazons and the Lanos of Mississippi.

The Mountain Range marked on the maps of Africa, under the title of Gebel el Kumir, or Mountains of the Moon, is, I believe, little better than-imaginary; if such a range exists, it may perhaps be found to correspond to that which in South America appears to extend from the sources of the rivers Paraguay and Parana to Minas de la Baheia and Pernambuco.

I lay little stress on these analogies; I mention them with a view to enquiry, not to conviction ; they are perhaps incorrect ; they are from their nature indefinite ; and I am fully aware, that whenever our ideas cease to be definite, they are apt to be fanciful.

——

The Direction which the waters of the Deluge observed in any particular district, may be determined by those who will employ the same diligence in exploring it, which Sir J. Hall has bestowed on the neighbourhood of Edinburgh : The method of doing this, is to examine the direction of bowlder-stones, mountains, valleys, promontories, and escarpments.

In regard to bowlder-stones enough has been said already ; I will only observe here, that blocks from the Cumbrian mountains have travelled eastwards as far as Pierce Bridge, and southwards as far as Staffordshire. Chalk-Flints occur in alluvion, on the north coast of Cornwall, and even at the Land's-end. Mr. Smith supposes the inundation which swept over England, to have come from the south-east, and though I am not aware of his reasons, I have no doubt that many may be found in support of that opinion.

———

The course of Mountain Ridges is as yet very imperfectly understood. Marsigli, Buache, Lehman, carried them uninterruptedly through the depths of the ocean. Continuous Ridges, like theirs, do not exist. As the banks of smaller vallies are intersected by the coombes and dales that fall into them, similar but larger and more frequent interruptions occur in the banks of those vallies which are more considerable. The Cevennes are divided from the Pyrenees by the plain in which runs the canal of Languedoc; from the Vosges, by a plain, which, extending from the Rhine to the Rhone, bifurcates near Langres. The valley of the Danube divides the Carpathian mountains from the Alps; the Riesen and Erzgebirge are detached from both : the mountain Chain of Wales is separated from

the Ocrynian by the Bristol channel ; from
the Cumbrian ª by a bay of the Irish sea.
The Scandinavian mountains are bounded
either by low plains, or by the sea.

What is a Chain of mountains ? An elevated platform, upon which rest various summits of unequal height. Wherever the
ground falls to a lower level than our platform, be its elevation what we please, there
the chain or ridge ends. If we place its
level above highwater mark, whatever lies
beneath high-water mark can form no part
of the Chain, and consequently the Chain
cannot be followed across a sea. If we

ª To the mountainous district of Cumberland, Lancashire, and Westmoreland, no name has yet been ap-
propriated. As it occupies nearly the extent of country
formerly inhabited by the Cumbri, I have ventured, on
the suggestion of my friend, the Rev. Wm. Conybeare,
to designate it by this term, which is short and easily
understood. I am indebted to the same gentleman for
the term Ocrynian, as denoting the high granitic tracts
of Devonshire and Cornwall. It is used in this sense by
Richard of Cirencester, and I have thought it better to
revive an obsolete name than to construct a new one.

place its level below high-water mark, whatever is dry land must form a part of the Chain, and consequently the Chain cannot be followed across a continent.

Buffon made the principal Chain in the Western Hemisphere run from north to south; in the Eastern from east to west; but corrected himself in the supplement to his work, where he states, that in both Hemispheres the principal Chains run north and south, and that those which run east and west are only subordinate.

Gatterer supposes different Chains, runing in different latitudes, to cross at intervals forming a kind of net-work. According to Pallas, they radiate from a common centre. His anonymous critic takes two principal Chains parallel with the equator, the one about $50°$ north latitude, the other about $25°$ south latitude, and supposes branches to be sent off from each of these towards the equator, and towards the poles.

Our knowledge upon this subject, confined and inaccurate as it is, enables us to pronounce that all these systems are erroneous.

Some modern writers have been disposed to confound chains of mountains with water-heads, and imagine that a line connecting the sources of rivers all over the globe must faithfully represent the line of greatest elevation. If the banks of rivers were in all places equally raised above their channel, it would do so ; but the reverse happens continually. The environs of Prague, I believe, are at a lower level than those at Töplitz, where the Elbe effects its passage between the Erz and Riesen-gebirge. The low plain, in which Strasburg and Manheim are situate, cannot compare in height with the ridges of the Hundsruck and Westerwald. The country about Ratisbon is by no means so elevated as that between Linz and Passau; following the course of the Danube from thence, its banks are at a low level till you reach the neighbourhood of Presburg, where it escapes through an opening left between the Alps and the Carpathian mountains. In Hungary it traverses a low plain. In Illyria it is again hemmed in by mountains.[a]

The Apalachian chain is notorious for the number of rivers by which it is intersected: equally celebrated are the gorges of the Potowmack and the Irtish.

Almost all the great rivers in Russia have their origin in low plains and morasses: there is no perceptible ridge along the water-head from Limberg to Petersburg, from Petersburg to the Oural.

This doctrine, therefore, cannot be relied upon; if it could, a ridge should be traceable from Weymouth to Fort William, from Bantry Bay to Colrain.

It has been thought that the highest mountains occupied the middle of conti-

[a] From Landrecie to Namur, the course of the Sambre is from S.W to N.E. the general slope of the country through which it runs, being from N.E. to S.W.: similar examples might be found in the Ardennes.

nents. The highest mountains[a], however, those of Mexico and Thibet, are both at an inconsiderable distance from the sea.

———

It has been said that mountain chains run down the middle of peninsulas : this doctrine is another instance of hasty generalization. The greatest peninsula, that of South America, is flat on the East side, and on the West, mountainous to the very edge of the sea; the structure of Scandinavia and India is the same.

It has been thought[b] that the steep sides of mountains front the West, but the reasoning by which this proposition is supported is lax in the extreme. In England and Scandinavia the steep side of the principal ridges front the N. W.: In Italy and Dalmatia the S. W.: so that the steep side of the former is at right angles with

[a] Journal des Mines, tom. xxiv. p. 303, and 352.
[b] Dr. Stukeley, who lived in the early part of the last century, is, I believe, the first author in whose works this remark is found; see Itinerarium Curiosum, Lond. 1724, p. 3. " If we cast our eyes upon the geography of Eng-

the latter, yet both have been said to face the West.

The principal chains of Europe and Asia, *viz.* the Alps, the Pyrenees, Carpacks, Erzgebirge, the ridges of Hæmus, Taurus, Imaus, seem to have their steep sides on the South. John Reinhold Foster, well known for the extent of his travels, conceived that in general the steep side of mountains was on the South and South-east. Bergman stated, that in mountains that run North and South, the West side

---

" land, we must observe that much of the Eastern shore
" is flat low ground, while the Western is steep and
" rocky. This holds generally true throughout the
" globe, as to its great parts, continents, or islands, and
" likewise particularly as to its little ones, mountains, and
" plains. I mean that mountains are steep and abrupt
" to the West, especially to the North-west, and have
" a gentle declivity Eastward, or to the South-east, and
" that plains ever descend Eastward. I wonder
" very much that this remark has never been made."
The subject is treated at considerable length by
Tilas, in an Essay published in the Memoirs of Stockholm, vol. xxii. for 1760, and by Cronstadt, in vol. xxv.
of the same work. Buffon recalled it to the attention of
geologists in 1778, and Jones in 1781; see also Walker
in Philosophical Magazine, vol. xxxv. and Kirwan
in Nicholson's Journal, 1803.

is steepest; in those that run East and
West the South.

A single glance of a good map of the
mountains throughout the world, if such
were to be found, would convince us that
all these hypotheses must be abandoned.

Promontories, like mountains and escarpments, may be found in every direction:
those of Cornwall, Carnarvonshire, Kerry,
Cape St. Vincent, Cape Verd, &c. stretch
Westward; in Labrador, Pernambuco, &c.
they stretch Eastward; but, viewing nature on a large scale, we shall perceive,
that as the great ridges, those at least of
the old continent, run East and West, so
the larger promontories in both tend to
North and South. Had their form been
determined by actually existing currents,
they would have tapered away to the West.
Their Southern Direction so strikingly exemplified in the three great peninsulas of
South America, Africa, and India, first re-

M 4

marked by Bacon[a], has been insisted upon
by Buffon and others. California, Alaschka,
Greenland, Kamschatka, Scandinavia, Florida, Italy, Greece, Arcadia, Arabia[b], the
Corea, are minor instances of promontories having the same direction. We are
prevented by the severity of the climate
from making ourselves acquainted with the
form of the Northern Extremities of Greenland and America : it appears, however, by
no means improbable that they may terminate in promontories directed North-
wards ; as is the case with Labrador, New-
foundland, Nova Zembla, Jutan, Jutland,
the Samoids, the lands at the mouth of
the Obi, and White Sea, Great Britain,
&c.

Have not Spitzbergen, the Norwegian
chain, that of Ural and Nova Zembla, the
appearance of being remnants of a promontory which stretched to the North?

[a] Bacon's Novum Organon, lib. i. Aph. xxvii. and
Opp. vol. ii. p. 8.
[b] Journal de Physique, vol. lxxi. La Metherie,
Leçons de Geologie, vol. i. p. 224. Bailly.

It has been supposed [a] that islands were
particularly numerous on the east of continents ; unfrequent on the west ; but the
exceptions to this rule are too many to
allow us to deduce from it any general
conclusion. Great Britain, the Isle of Man,
the Orkneys, Shetlands, Hebrides, the
Scilly Islands, the Canaries, Madeiras,
Azores, are all situate on the West.

To determine the Age of the world has
long been a favourite object with philoso-
phers. Halley, having persuaded himself that
the sea increased in saltness, suggested, as
a mode of solving this problem, an examination of the quantity of salt contained
in a given portion of sea-water, in distant
periods of time. Ricupero counted the
beds of lava upon Etna, and, from the average of time which he supposed to intervene between the several eruptions, under-
took to calculate the age of that mountain,

[a] Bacon's Novum Organon, lib. i. Aph. xxvii. and
Op. vol. ii. p. 8.

and by analogy, the age of the earth. The disintegration of rocks, the mouldering of hills, and the gradual filling up of valleys, by the debris which falls into them, were adduced by Burnet, as conclusive arguments against the high antiquity ascribed to the earth by the writers of that day. Deluc, Dolomieu, and Cuvier have distinguished themselves by the attention they have bestowed on other instances of diurnal change. After a patient investigation of the phœnomena of bays, promontories, deltas, dunes, taluses, seas, lakes, and rivers, they are agreed in thinking that the period of time, which has elapsed since the retreat of the diluvian waters, cannot exceed from five to six thousand years.

So much for the positive age of our planet ; — let us now consider its age in relation to different events connected with its own history, the history of the solar system, and the history of mankind.

That the order of things, as it existed

before the deluge, cannot have differed widely from the present order, will appear from many considerations.

1. The earth having acquired a spheroidal figure, while fluid, must have revolved even then upon its axis of fixed rotation ; now it is extremely improbable that the earth should have had this motion at a time when the sun and planets were not yet called into existence.

2. In the diluvian detritus of almost every country which has been examined, have been discovered bones of the horse, ox, stag, elephant, and other quadrupeds. These animals inhabited the earth ; consequently they had land to roam on, plants to feed on ; the animals and plants grew and flourished : consequently they must have enjoyed an atmosphere and a climate suited to their nature; in other words, an atmosphere and a climate varying little from those of the present world. If so, may we not conclude that the antidiluvian earth was a planet belonging to the solar system, re-

volving in an orbit little different from its
present orbit, and undergoing all those periodical changes, upon which climate, temperature, vegetation, and animal life depend?

3. The several planets are spheroidal
like the earth; therefore they have been
fluid: and they agree with the earth in so
many other particulars, that physical astronomers do not hesitate in ascribing to both
a common origin.

Dr. Herschel deduces from his observations on Nebulæ, that they consist of rare
and luminous matter, gradually condensed
in consequence of the attraction of denser
nuclei which they surround; if we may
suppose the heavenly bodies formed by the
same process, Comets, he thinks, would afford an example of an imperfect, Planets
of a complete condensation of such matter.

The author of the Mechanique Celeste,
has with becoming caution advanced a similar hypothesis; he supposes that the matter
of the solar atmosphere expanded by exces-

sive heat to the limits of the solar system, revolved formerly round the centre of gravity of the system, and that this matter becoming condensed by cooling, the greater portion of it was attracted to the sun, its centre; but smaller portions to as many other centres as there are planetary bodies : the zodaical light he attributes to the remaining portions of the same atmosphere still uncondensed and revolving round the sun.

By this hypothesis, La Place is enabled to account for the former fluidity, and present spheroidal figure of the planets, for the small eccentricity of their orbits ; and for the motions of the sun, the planets, and their satellites upon their axis of rotation, and of the planets and their satellites in their orbits being in the same direction, and nearly in the same plane; for the bodies thus formed, will necessarily retain the motion of the atmosphere in the plane of its equator, and the exterior zones of atmosphere having a greater absolute velocity than the interior, the bodies formed by the condensation of such zones, will have a rotatory motion in the direction of the motion of the atmosphere itself.

The extent of the operations which we have here contemplated, is so vast as to be embraced with difficulty by the most capacious mind.

Long accustomed to admire that uniformity of movement which the lapse of five thousand years has been unable to disturb, having continually more and more reason to believe that the solar system contains no seeds of decay, and that, as far as the motions of the sun and planets depend only on their mutual action, that system may have subsisted from, may endure to eternity, we are naturally slow to admit that the world has ever existed under circumstances so different from the present as these theories suppose: but experience cannot furnish a clue to the history of times extremely remote from those in which human experience has been collected; and the spheroidal form of the earth is, of itself, sufficient to convince us, that the course of

nature has not always been uniform, that
her laws are not absolutely fixed, and that
the solar system, unchangeable as it now
appears, has, notwithstanding, had a beginning, and may have an end.

This opinion is further supported by
analogy. Within the short period of astro-
nomical observations, changes are known
to have taken place in some of the heavenly
bodies, changes no less extraordinary than
those which La Place and Herschell imagine, at an earlier period, to have affected
the earth. As an instance of this, we may
mention a star in Cassiopœa : first observed
in 1572, it gradually acquired a brilliancy
which exceeded that of Sirius, so that it
could be seen during the day ; but this brilliancy soon declined, and in 1574 the star
was no longer visible ; and yet, in all probability, it continues to exist in the region
where it was last observed, though in a
state of opacity ; for its change of splen-
dour was not attended with any change of
place.

" How prodigious a change" exclaims

La Place, " must this vast body have un-
" dergone! how far do such operations
" surpass any which our sun presents!
" how clearly do they prove to us that
" nature is far from remaining always and
" every where the same!"

Whatever weight the reader may be disposed to give to the hypothesis above
quoted, I think he will be disposed to admit all I wish him to admit at present, *viz.*:
That the several planets acquired their
positions, their spheroidal forms, their
fixed axes of rotation, their velocities and
the common direction with which they
move upon their axes and in their orbits
round the sun, at the same time and by the
operation of the same cause; and consequently that the diluvian catastrophe did
not take place till after the establishment
of the solar system.

———

That the earth was divided into land and
water, at a period antecedent to the deluge,

is evident, from the remains of land and sea productions so abundantly diffused throughout the secondary rocks ; but the situation which the land and sea respectively occupied before this event, appears in many instances to have differed materially from that which has been since assigned to them. This circumstance, which we might have reasonably anticipated, on considering the changes of form which the surface of the earth must necessarily have undergone, from the excavation of rock in some places, and the accumulation of detritus in others, appears indisputable when we contemplate the situation of the different places to which bowlder stones and gravel have been transported.

The blocks of granite on the Jura attest the non-existence of the Lake of Geneva, at the time of their transportation.

The similarity of the parasitic gravel and soil at Malta and Gozo attests the non-existence of the Mediterranean at the time the gravel and soil arrived at these islands.

N

The blocks of primitive Norwegian rocks scattered over the North of Germany, Russia, Holland, and occasionally met with on the eastern coast of England, attest the non-existence of the Baltic and German Sea, while these blocks were in motion.

Pieces of granite are found on Staffa, which could only have been brought thither when Staffa was annexed to the main-land ; and similar phenomena will probably be found in every other part of the world, when they become objects of inquiry.

The newest formations with which we are acquainted are intersected by valleys, and covered by alluvial deposits : hence it follows that this event was posterior to the birth of these formations.

Inattention to this circumstance has occasioned many errors in Geology. I remember being told by a Professor on the Continent, that the frequent change of texture, colour, and grain, observable in the granitic rocks in the neighbourhood of Heidelberg, was owing, he conceived, to the depositions having been disturbed by the great valleys of the Rhine and the Neckar. Though Lamanon thought that currents were rather the effect of valleys than the cause of them, even He, has not always distinguished the process by which strata were formed from that by which they were mutilated ; and Dolomieu[a] furnishes, in one of his best memoirs, an example of that unfortunate association of order and confusion, which, for want of correct notions on the subject, he has ventured to ascribe to nature.

If this event was posterior to the consolidation of the most recent rocks, it was obviously posterior also to the interment of the fossil organic bodies which these rocks contain.

Woodward, Scheuchzer, Buttner, Lehman, and the pupils of Hutchinson, attributed fossil shells to the deluge, as the

[a] Journal de Physique, vol. xxxix. p. 391.

common people generally do at this day.
Stukely[a] also fell into this very natural
error. " If," says he, " we observe, how
" the Lincolnshire Alps run fifty miles
" north and south, and on the west are
" steep and rocky, we see why the strata
" near Newark are so stocked with shells ;
" for it is reasonable to suppose, that on
" the retiring of the waters of the deluge
" from the superficies of this country into
" the eastern seas, these heavy bodies
" were intercepted by this cliff, which has
" retained such vast quantities of them ever
" since, while those that fell on common
" mould are mostly rotten and now lost."

Targioni[b], Arduino, Rouelle[c], Hollman [d],

[a] Transactions of the Royal Society for 1719.
[b] Targioni and Arduino maintained that there had
been many deluges. See Breislack's Introduction to
Geology, p. 376.
[c] Rouelle shewed that fossil shells do not lie at ran-
dom ; that different kinds of shells are found in different places ; and that they occur in different strata.
[d] In an elaborate treatise published in 1772, in the
Introduction to the Journal de Physique, tom. ii.

Gesner, Buffon, Whitehurst [a], &c. have
succeeded in some measure in correcting
this error: but many of our cotempo-
raries are, I fear, in danger of falling,
some of them, indeed, have actually fallen,
into a mistake no less grievous ; that of
attributing the productions of the Sub-
Appennines [b] to the Adriatic, and those of
Nice [c] to the united waters of the Euxine
and Caspian forcing a way through the Hellespont into the Mediterranean.

I am at a loss to determine by whom it
was first observed, that fossil shells had
their nearest analogies in climates different
from those in which they are now found.

---

p. 118, Hollman contends that fossils are not owing
to the deluge, a doctrine advanced shortly afterwards by
Gesner (Journal de Physique, Introduction, tom. ii.
p. 608. but with some modifications.)

[a] Whitehurst (Theory of the Earth, p. 59.) says, that
beds of fossil shells which consist of one species only,
and are not native to the climate where found, but of
very distant regions of the earth, show that the inhabitants of these shells have lived and died in the beds in
which they are found.

[b] Brocchi Conchyologia Sub-Appennina.

[c] Rizzo. Journal de Physique.

The circumstance was well known to le Maillet[a], Jones[b], and Catcott.[c]

In the petrifactions of Monte Bolca, where the impressions of fish are preserved between the laminæ of a calcareous schistus, a hundred and five different species have been enumerated ; of which thirty-nine are said to have come from the Asiatic seas, three from the African, eighteen from those of South, and eleven from those of North America. It is a saying which, I presume, the observations of M. Cuvier will not warrant ; but admitting the fact, are we to believe that Monte Bolca has been situate in different and distant places at one and the same time ? that all these various animals repaired to this spot, (the area of which does not exceed that of the Banqueting-house at Whitehall,) from seas occupying the opposite quarters of the globe ? seas, which, as has been shewn, did not exist till after their supposed inhabitants had emigrated and pe-

[a] Telliamed.
[b] Physiological Disquisitions.
[c] On the Deluge, p. 251.

rished? or is it not more reasonable to suppose, in compliance with the theory here advanced, that there is no connexion between these recent bodies and the fossil ones, except that of resemblance?

Jussieu [a] pronounced the plants of which impressions are met with in the coal mines of France, to be tropical. The resemblance is admitted; but here again I ask, shall we believe that France once possessed the loco-motivity piously ascribed to the Chapel at Loretto? or that these plants, growing on a soil not yet in existence, floated with impunity over the not yet existing Atlantic? or, rejecting both these suppositions as contrary to common sense, shall we not believe, that the resemblance between the impressions of plants found in the Coal Mines of France

[a] Even Mr. Playfair conceives that these proofs of the transportation of materials " *by the sea*," have the advantage of involving nothing hypothetical, and that the accurate comparison of the animal exuviæ of the mineral kingdom with their living archetypes, may lead to important consequences concerning the nature and direction of the forces which have changed, and " *are continually* changing" the surface of the earth. See Illustrations, p. 178.

and those now growing between the tropics, is merely accidental? that they are productions of a similar climate if you please, but not of the same world?

Alarmed by the conclusions which necessarily flow from such premises, as that our northern strata owe their fossil productions to southern climates, many naturalists have of late attempted to prove, that the fossils in question, though reputed tropical, do in fact exist in seas nearer home; but, though they should succeed in this attempt, which I doubt, the chief difficulty will still remain; for let it be assumed that the archetypes of shells found in France, are those of the Atlantic; in Italy, those of the Adriatic and Mediterranean; still we might as well suppose, that Hannibal obtained his vinegar from a modern commissariat, as suppose, that the fossil shells of France and Italy were derived from the seas in their vicinity, before those seas were in existence.

———

That the Deluge in question was posterior to the birth of mineral veins, and many, if not all, basaltic dykes, is deducible from the intersection of these veins and dykes by valleys, and the occurrence of their detritus in stream works.

Werner [a] tells us that the occurrence of veins depends much on the external form of mountains.

1. On the position of the whole chain of mountains in respect to its extent and declivity.

2. On the particular position of the country where they occur.

Whether the country be composed of hills with gentle declivities and roundish or flattish summits, or

Whether it be a place in a principal valley.

Those who can unravel the meaning of this passage will find the ideas contained in it erroneous.

He supposes also, that the fissures in which metallic ores occur are often occasioned by mountains having fallen to-

[a] Werner on Veins, Translation, p. 54.

wards the sea or valley; if so, we have shown, that such sea or valley must have existed before any which now exist.

We have no positive evidence to determine whether the deluge took place before or after the Creation of Man : we have only this negative evidence, that neither any part of a human skeleton nor any implements of art have been hitherto discovered, either in regular strata, or in diluvian attritus.

As for human bones, M. Cuvier [a] says, " it is certain that none have been met with " among fossils properly so called. Our " workmen about Paris almost all believe " the bones so frequently found in the gyp- " sum quarries to be human; but having seen " several thousands of these bones, I may " be permitted to say, that not one of them " has ever belonged to our species. At Pa- " via I examined a number of bones which " had been brought from the island of Ce- " rigo by Spalanzani, and, in spight of

[a] Recherches, Disc. Prelim.

" what that celebrated observer has said of
" them, I pronounce that not one among
" them is human.

" The fossil which Scheuchzer called
" *homo diluvii testis*, I have restored to its
" proper place among the Salamanders.

" At Canstadt in Franconia, a fragment
" of a human jaw was found, but we know
" not at what depth, or under what cir-
" cumstances.

" Every where else, the bones supposed
" to be human turn out to be those of some
" other animal."

Since this passage was written, human
skeletons, imbedded in stone, have been
found in Guadaloupe. Mr. Konig [a] has published an account of the most perfect, I believe the only one, of these that has been
brought to Europe ; in all probability the
stone is a recent concretion of calcareous
sand on the sea shore.

At the convent of Rossweil, in Swisserland, I was shewn a great curiosity, preserved in a shagreen case, richly carved
and gilt on the outside, and lined with

[a] Philosophical Transactions, 1814.

velvet. The monks called by the name of an antedeluvian knife, a piece of limestone accidentally broken into a form somewhat resembling that instrument. Dr. Hook[a] gives several instances of ships having been found in mines and in the bowels of the earth.

It is said by Mr. Knight Spencer[b], that an ancient brass pin has been found in the heart of a flint ; and an instance is adduced in the Journal de Physique[c], of copper nails having been discovered in limestone.

In the Journal des Mines[d], money is reported to have been seen in flint. Lamanon[e] mentions a key found in the interior of a block of gypsum at Montmartre, and La Metherie

[a] Posthum. Works, p. 439. 441. 443. See White-hurst, p. 133.

[b] Bakewell's Introduction to Geology, p. 338.

[c] Tom. xxxi. Supp. p. 70.

[d] Tom. iv. p. 76.

[e] Journal de Physique, tom. xvi. See also on this subject, Linnæus, Wormius, Grew, Zannechelli, Henck-el, Scheuchzer. Journal de Physique, for 1772. In-trod. vol. ii. p. 549. Spalanzani, in Journal de Physique, vol. xlvii. p. 281. and vol. lxi. p. 51. vol. xxvii. p. 168. and vol. xiv. p. 302. Jacob's Travels in Spain. Guet-tard's New Memoirs, vol. ii. p. 314.

speaks of a horse-shoe, found under similar circumstances on that hill.

Joinville and de Saade, I think, have recorded an observation of the same kind which they made at Aix.

All these accounts are, however, too apocryphal to be admitted without further proof of their authenticity.

Of the bones which occur in diluvian gravel, some are analogous to those of species now existing; of others even the genera are unknown to us. What shall we say then? that man, the monarch of creation, was once the contemporary of the mammoth? or that the elephant, the horse, the pig, are of a more ancient family than an Howard and Montmorenci?

Thus have we been able to point out with some degree of confidence, the relative æra at which the deluge took place; but if we would proceed further, if we would investigate the means by which this tremendous catastrophe was produced, the mind is easily bewildered in unprofitable conjecture.

If the submersion of the highest mountains on the face of the globe was occasioned simply by an Increase of Water,
from what source can so enormous an addition of water have proceeded? If it
existed previously, what became of it during the growth of those land-plants, which
we so often find imbedded in the secondary rocks? during the life-time of
those land-animals whose fossil remains
are so extensively distributed? if it existed at the time of the Deluge, what is
become of it now? if derived from the interior of the Earth, as Sir H. Englefield
supposes, (a supposition not easily recon-
cileable to what little we know of its interior from the experiments of Maskelyne
and Cavendish,) how explain the existence
of those enormous caverns, within which this
mass of water was contained? how explain
its own existence in such a situation? what
attraction from without, what repulsion
from within could have dislodged it from its
hiding place, and forced it far beyond
those barriers which the laws of gravity
prescribe? how happened it that the roof
and sides of the caverns, in which the water

resided, did not fall in during its absence, so as to prevent the possibility of its return? Was Increase of Temperature the means of dislodging it? whence did that increase of Temperature proceed? from within? we know not any cause acting from within capable of producing it; of producing it once, and once only, within a space of five thousand years: from without? how could heat be at the same time so intense as to penetrate a solid crust some thousand miles in thickness, and yet so gentle, that no traces of its action are discerned upon the surface, where it must have acted most intensely?

If it be supposed that this accession of water was derived from some body extrinsic to the earth, we know of no cause in nature by which such transfer of water from one body to another could be produced: but let a cause be assumed; let us grant that the water was so obtained; how was it afterwards removed? what is become of it now?

Shall we then, fearless of paradox, attribute to the waves constancy, mobility to the land? Shall we say that continents have been submerged, not from the rising of waters, but from their own descent? Extravagant as such an hypothesis may appear, it falls short, very short of that which the Huttonians have long admitted and maintained. " There can be no doubt," says Mr. Playfair, " that the land has been " raised by expansive forces acting from be- " low; and there is reason to think that con- " tinents have alternately ascended and de- " scended within a period comparatively " of no great extent."

Under the sanction of such authority, may we not hazard the moderated theory, that, once, and once only, continents have stooped from their elevated station, in order to shield ourselves from the consequences, with which we are threatened by giving way to the swollen pride of the ocean?

Alas! this expedient, so far from obviating our difficulties, tends only to enhance them.

If there were no caverns beneath our continents, how could they sink?

If there were caverns, how were they produced? why were they commensurate with the extent of the land?

The continents having sunk, how have they risen again to their present level?

After all this subsidence and elevation, how happens it that of the strata which were deposited horizontally so many remain horizontal?

How happens it that subsidence and elevation were unattended by fracture?

But the submersion of the earth is not the only condition required to bring about a state of things such as we have described. Valleys could never have been excavated, nor huge bowlder-stones have been transported to so great a distance under water, had the water been subject to such comparatively trifling agitations only, as those by which that fluid is affected in the present constitution of the world.

To the solution of the problem Impetuosity of Motion in the water is indispensible; but an increased Quantity of water is, perhaps, superfluous; for there seems no good reason for supposing, that the quantity which actually subsists upon the earth, if thrown into a state of excessive agitation, would not be of itself sufficient to produce all the phenomena of the deluge.

We have seen that previously to that catastrophe, the general state of things upon the earth was very much the same as at present; that there existed land and sea, both of them inhabited; that the earth was a planet revolving on the same axis as now, warmed by the same sun as now, and nearly to the same degree. We have seen also, that this order of things, so closely resembling the present order, was suddenly interrupted by a general flood, which swept the quadrupeds from the continents, tore up the solid strata, and reduced the surface to a state of ruin: but this disorder was of short duration; the mutilated earth did not cease to be a planet; animals and plants, similar to those which had perished, once more adorned its surface, and nature again submitted to that regular system of laws, which has continued uninterruptedly to the present day.

Where then was the cause of this transitory but tremendous disturbance?

We are not aware of any force depending

on the internal Constitution of the Earth, that could now effect so great a revolution as the deluge ; therefore, it is not probable that the deluge was effected by a force residing *within* it, immediately before the deluge ; for the constitution of the earth was at that period nearly the same as it is now.

Did the disturbing cause reside then in the mechanism of the Solar System ? No ; our knowledge of the laws which regulate the motions of the planetary bodies, aided by an experience of five thousand years, will not allow us to admit that this system contains any seeds either of derangement or decay.

It must have resided, therefore, *without* that system.

If we enquire the *extent* of the disturbance, it modified the outward form of the earth, but without affecting its interior constitution, or exerting beyond the confines of the earth any influence with which we are acquainted. The order of things, which subsisted immediately after the deluge, so much resembled the order of things which subsisted immediately before it, as to preclude the supposition that the earth, when considered in the character of a planet, underwent during that eventful crisis any material revolution ; such a revolution it must have experienced, if the force acting upon it had been either the cause or the effect of a change of motion or position in any other member of the solar system.

If then we would discover the Cause of this catastrophe, we must look for a Cause foreign to our globe, foreign to the solar system, capable of inundating continents, and giving to the waters of the deep unexampled impetuosity, but without altering the interior constitution of the earth, or deranging the sister planets ; moreover the Cause must be transitory, and one which, having acted its part once, may not have had occasion to repeat it in the long period of five thousand years. Any supposeable Cause that would not fulfil these conditions, is insufficient for our purpose.

Would a Comet fulfil them? Much

would depend on its bulk and distance. It would not fulfil them if we suppose a Comet, large in comparison of the earth, to move in a line joining the centres of the two bodies, so as to produce a direct shock; but, if we suppose one of suitable dimensions to move in such a direction as would allow it only to graze the earth, it is not impossible that the shock of this body, a body, such as we require, out of the solar system, might produce the degree and kind of derangement which we are attempting to account for; I mean a great temporary derangement on the surface of the earth, unaccompanied by any material change of its planetary motion. Euler, who, in a treatise entitled " *De periculo a nimiâ cometæ* " *appropinquatione metuendo*," has investigated the changes that would be made in the elements of the earth's orbit by a Comet, its equal in bulk, coming almost in contact with it, finds that the attraction of such a Comet would indeed alter the length of our year, but only by the addition of seven hours. The maximum effect resulting from the Comet's attraction at the time of its passage, would be greater than we should be led to infer from the total result of its attraction, after its final departure ; for the changes occasioned during its approach, would be in a great measure undone during its retreat ; but even at their maximum they would not be very great, because from the rapidity of the Comet's motion, time would be wanting to complete them. A Comet grazing the earth, would be incompetent, Euler says, to produce even a deluge of our continents, unless the shortness of its stay were compensated by a magnitude of volume, exceeding that upon which he has founded his calculation.

I shall conclude by remarking, that if the hypothesis of a shock derived from the passage either of a Comet or of one of those numerous, important, and long neglected bodies, often of great magnitude and velocity, which occasion meteors, and shower down stones upon the earth, would explain the phenomena of the deluge, (a point upon which I forbear to give any opinion,) we need not be deterred from embracing that hypothesis under an apprehension that there is in it anything extravagant or absurd. In the limited period of a few centuries, there is little probability of the interference of two bodies so small in comparison with the immensity of space; but the number of these bodies is extremely great, and it is therefore by no means improbable, says La Place, that such interference should take place in a vast number of years.

ESSAY III.

ON THE INEQUALITIES WHICH EXISTED ON THE SURFACE OF THE EARTH PREVIOUSLY TO DILUVIAN ACTION, AND ON THE CAUSES OF THESE INEQUALITIES.

I HAVE endeavoured in the preceding Essay to refer to their immediate causes, the inequalities which diversify the present surface of the earth ; but, as the operation of those causes must have been greatly modified by the form of the surface on which they acted, it is necessary, in order to complete our enquiry, that we should endeavour to ascertain the figure of the earth in preceding ages.

While its construction was yet going on, is it probable that the surface of our planet exhibited one uninterrupted plane, or that it was diversified then, as it is now, by protuberances and depressions?

In favor of the former opinion, may be cited the authority of Stracey, Hutchinson, and many of the early writers. Believing that the materials which constitute the solid crust of the globe, were deposited from a fluid menstruum, in obedience to the laws of gravity uninfluenced by disturbing causes, they inferred, *à priori*, that every stratum must have been originally plane or rather concentric. The observations of Dr. Richardson [a] go to justify that inference : for though the upper surface of superficial strata is continually scolloped away, the plane forming their base, he says, continues always steady and rectilinear, and the upper and under surfaces of those beneath he finds invariably parallel to one another.

These observations, however, I apprehend, are just only when applied to districts of inconsiderable extent. Even along the coast of Antrim, it would not be easy to find the surface of the chalk unlacerated for a mile together, protected as it is by its thick cover of whinstone.

The Blackheath gravel bed is made up of chalk flints; the chalk must therefore have been partially destroyed before this bed was formed.

Can the chalk of England be unlacerated where covered by a stratum of Woolwich pebbles, pebbles which consist of rounded flints broken from the chalk beds?

Mountain chains, composed of primitive rocks, are often separated by valleys in which the rocks are secondary. Between the Ocrynian ridge in Cornwall and Devonshire, and the mountains of Wales, red sandstone, mountain limestone, coal-measures, &c. unconformably disposed, occupy a valley, through which has since been excavated, by diluvian action, the interior vale of the Severn; similar intervals between the Welsh and Cumbrian mountains, between these and the Cheviots, between the Cheviots and the mountains of Scotland, exhibit similar appearances. How could these hollows and ridges be so occupied, if protected strata were always even, or if the surface of the primitive strata had not been uneven in these places at the time when the secondary strata were deposited? Granite, commonly said to be the basis on which all other rocks repose, is found on the highest eminences, at the lowest depths with which we are acquainted. Coal is often represented as occurring in basins or troughs. Who has not heard of the upfilling of strata? these are additional instances of irregularity of surface in protected strata.

Many of the secondary rocks contain pieces of wood or impressions of fern or fossil shells. Unless we suppose that marine moluscæ inhabited the land, or timber grew upon the ocean, we must admit that there existed, at the time when these were deposited, land and water; in other words, mountain and valley.

Inequalities of surface prevailed then,

even during the formation of strata ; to what causes are these inequalities to be referred?

1. Crystallization.

This term is employed in different senses. It sometimes means an aggregation of integrant molecules arranged by polarity into a definite essential form, each molecule being composed of the same elements united in the same proportions. Professor Jameson appears to think that even this, the most perfect species of crystallization, has not been inactive in determining the figure of the surface of the earth : what are called strata he supposes to be in many cases mere laminæ of crystals, which, being produced, would be found to meet at determinate angles. I do not know the grounds on which this opinion rests ; but strata and beds agree so closely in character, that I cannot persuade myself they owe their existence to different causes. Now there is no instance known of a crystal made up of dissimilar laminæ. The primitive molecule of many of the metals is the same, notwithstanding which we have never seen a crystal composed of layers of different metals ; for instance, of red copper ore, galena, and iron pyrites ; still less a crystal composed of alternate layers of substances, the primitive molecules of which are dissimilar, as alum and sulphur ; how then can it be imagined that mountains composed of different strata, formed at different periods, are crystals ?

By irregular crystallization, I understand a concurrence of similar or even dissimilar integrant molecules associated without regard to polarity. This prevails in all rocks which are foliated, radiated, or fibrous, and in many consisting of angular grains. The ingredients of these rocks, on being disengaged from the fluid which held them in solution, are thought by Rouelle, and his pupil Lametherie [a], to have disposed themselves in obedience to the laws of elective attraction, in groups insulated at their sum-

[a] Journal de Physique, vol. xlii. p. 132. 294. and 445.

mits, united at their base. These groups it is said, formed original mountains, the intervals between them original valleys.

Their theory to this extent appears reasonable; but they have not stopped here: they have referred to irregular crystallization effects produced upon non-crystalline materials; upon brecchias, shell-limestones, &c. ; even the cliffs near Calais, according to Lametherie, are results of crystallization.

2. Partial deposition.

This would result from the different state of the menstruum in different places, occasioned principally by tides and currents, the influence of which must have been felt from the first moment that the earth and moon were in existence; for the quantity of matter deposited at each place would be inversely as the quantity of motion in the fluid from which it was disengaged.

The multifarious productions of secondary rocks afford ample testimony of these tides and currents. To what other cause can we attribute the frequent intermixture of animals inhabiting the land, with those inhabiting the sea? of wood, fern, bones of lacertæ, moluscæ dwelling only in shallows, with pentacrini dwelling only in the deeps?

If partial deposition tended, in some places, to increase subsisting inequalities of surface, in others it would tend to diminish them: it would tend to diminish them wherever the depositing fluid experienced interruption from ridges previously existing; that is, practically speaking, in all those situations where we find on one side of a mountain chain a different series of rocks from that which we find upon the other. [a]

Thus porphyry is very common on the Italian side of the Alps, rising between Bolzano and Brixen to the height of 4000 feet, but on the German side it is altoge-

[a] See Journal de Physique, tom. xlix. p. 212. See also Voyages de Saussure, § 981. and Nicholson's Journal, vol. iv. p. 264.

* o 8

ther wanting. On the German side, serpentine and other magnesian rocks abound, which are rarely met with on the Italian.

In the same manner the deposition of red marl, which occurs below the fells of Alston Moor, and the Cumbrian group, in the valley of Carlisle, does not appear to have ever extended beyond them.

3. Subsidence.

On the same principle as new houses settle, new rocks would subside, and this, not only in consequence of their own weight, but in consequence of the weight of other strata heaped upon them.

The degree to which this cause has operated may be ascertained; for every subsidence implies a fault, and consequently where there is no fault we may safely conclude there has been no subsidence.

It does not follow, however, conversely, that where there is a fault, there must have been subsidence; for, if new strata were now to be deposited, the faults would be as numerous as mountains and valleys, occasioned not by subsidence but inequality of surface.

The relative period at which subsidences have taken place, may often be determined by examination of the superior strata.

The nature of the substances by which chalk is covered at the Isle of Wight, precludes the idea that they could have been originally vertical; they are vertical now and parallel to the chalk; hence the subsidence of the chalk could not have taken place till after the deposition of the beds that cover it.

The unconformable position of the red marl in this country and in France, evinces that the coal and limestone had acquired their inclination before this marl was deposited. It is commonly said in England, that the coal is cut off by the red ground; it would be more correct to say the red ground is cut off by the coal.

The general constancy which prevails in the direction and dip of mountain chains, countenances the idea that many of these

P

chains, however modified by diluvian action, owe their origin to subsidence.

4. Volcanoes and Earthquakes.

It is probable that the effect produced upon the earth by these, was confined in the earliest ages as now, to forming occasional hills by the accumulation of ashes and scoria, or occasional valleys by the falling in of unsupported craters.

But the action of running water appears to have been in all ages the principal cause of inequality of surface.

In many places, at the Valorsine notoriously, conglomerate is found interstratified with beds, to which, but for this circumstance, no one would deny the appellation of primitive. Croagh Patrick, a mountain which ought to be attractive to Geologists as well as Pilgrims, is composed of quarz rock lying amid clay-slate, serpentine and mica-slate, which contains in some places large nodules of quarz.

Dr. MacCulloch has given, in the Geological Transactions, repeated instances of the same kind, which, being sanctioned by his own experience and that of continental observers, have induced Mr. Jameson to admit Quarz rock, Sandstone, Greywacke, and Conglomerate, into his list of primitive formations. I know — chemical products sometimes bear so striking a resemblance to mechanical, that it is not easy to distinguish them ; and it is improbable that the rocks just mentioned are made up only of abraded portions of others previously destroyed ; but few will deny that they contain some abraded matter ; and that abrasion can be accounted for only by the instrumentality of running water.

If the effects of running water are faintly traceable on the primitive rocks, they are very conspicuous in those of a later period. " A great catastrophe," says Dolomieu [a], " seems to have taken place after the birth " of the primitive rocks, and before that of " the derivative or parasitical (*couches de* " *transport*). Regularity of structure ceas- " ed ; a fracture took place in consequence

---

[a] Journal de Physique, tom. xxxix. p. 390.

" of some vast shock ; vast it must have
" been, to break through a compact shell
" 4000 fathoms in thickness. The strata,
" which precipitation had arranged hori-
" zontally, and crystallization consolidated,
" were thrown up ; some, vertically to a
" height which, since that period, the water
" has never attained ; others, obliquely in
" various directions : thus were formed the
" great eminences of our globe, from which
" were derived its present irregularities of
" surface."

Without assenting to every part of this doctrine, I cannot but consider the almost universal occurrence of conglomerate and greywacke on the confines of what are called primitive rocks, as one of the most important and striking facts yet established in Geology : it seems to prove, that, at the epoch at which these beds were formed, a deluge took place, similar in kind, though perhaps not equal in extent, to that which determined the present outline of the earth.

From that period till the work of creation was complete, the mechanical action of

running water appears to have been more
gentle and partial, but unremitted. The secondary beds are obviously composed, for
the most part, of the ruins of their predecessors ; their surfaces are continually water worn, and the alternate recurrence or
intermixture of various productions of the
land and of the sea, of fresh water and salt
water, so eloquently described by Cuvier
and Brongniard, is by no means confined to
the narrow district, to the small number of
beds, which forms the subject of their Essay
on the Mineralogical Geography of the
neighbourhood of Paris, but pervades, indiscriminately, almost every country in
which transition or secondary rocks are
found, and almost every member of those
formations.

ESSAY IV.

ON FORMATIONS.

By the term Formation is meant a series of similar or dissimilar rocks, supposed to have been formed in the same manner and at the same period.[a] The idea is therefore purely theoretical.

Two circumstances are thought to justify our assigning to different substances a common antiquity and origin; an inter-mixture of their ingredients, and alternate occurrence.

[a] In Journal des Mines, tom. xxvi. p. 170, we are told that strata of the same formation may be of different ages; but the definition here given conveys the sense in which the word formation is, I believe, most generally used.

An intermixture of ingredients however proves little more than contiguity. Rocks admitted to be of very different ages often exhibit it, while it is as often not discoverable in rocks thought to be of the same age.

Near Ravenstone, in Westmorland, the passage from mountain limestone into greywacke-slate is imperceptible: at a short distance from this spot the greywacke-slate, inclined at a high angle, is covered by conglomerate, and rounded pieces of slate are enveloped in the horizontal limestone above: shall we say of the same beds, in one part of their course, that they are members of the same formation, in another that they belong to different formations?

Who can distinguish the greywacke-slate from the red rab in the neighbourhood of Milford Haven, or in the south of Ireland? Who can say, in Herefordshire, where the old red ceases and the mountain limestone begins? What mineralogist can draw a line of demarcation between the red marl and toadstone at Heavitree? between the old and young red on the shores of the

Bristol Channel? between the young red
and mulatto along the vale of Geneva?
between mountain limestone and slate in
many parts of Devonshire? between limestone and lias in the district that stretches
from Auch to Cahors? between lias and
mulatto at Argenton? between mountain
limestone and chalk in the Italian Alps?
But will any one affirm that the rocks so
blending with each other in external character are of the same formation?

So in regard to alternate recurrence. —
Alternating substances are not always
coeval; nor do coeval substances always
alternate.

The limestones, which near Plymouth
alternate with slate, agree with those which
in Cumberland alternate with coal. Are
the slate and coal of the same formation?

Granite is of one formation; Gneiss of
another: Do these never alternate, or graduate into each other?

ON UNIVERSAL AND PARTIAL FORMATIONS.

Similar substances are occasionally found under similar circumstances in distant parts of the world. On this slight basis has been raised the imposing superstructure of universal formations.

Sir Robert Atkins [a] assumed, that if a line were drawn from the mouth of the Severn to Newcastle, and continued round the globe, coal would be found near that line, and scarce any where out of it. Werner's theory was more comprehensive [b]; he conceived that the greater part of the primitive, transition, and flötz formations were

[a] Strange's Phil. Transactions.

[b] It would seem from the following passages, that Woodward had an indistinct perception of the same theory.

" I was abundantly assured that the circumstances
" of these things in remote countries were much the
" same with ours here — that the stone and other ter-
" restrial matter in France, Flanders, Holland, Spain,
" Italy, Germany, Denmark, Norway, and Sweden,
" was distinguished into layers as it is in England, &c.
" — I got intelligence that these things were the same
" in Africa, Arabia, Persia, and other Asiatic provinces,

continued in the same manner round the
globe, though not uninterruptedly.

To form a correct opinion on this subject, we must descend from generals to
particulars.

If we consider the individual strata of
which a formation is composed, so far from
being able to trace these round the globe,
it is generally impossible to do so to the
extent of a few miles; for the regular order
of succession, as it is called, is perpetually
disturbed, either by the interposition of a
new substance, or the discontinuance of an
old one, or the substitution of one substance for another, or the splitting of one
stratum into many.

Chert, flint, septaria, gypsum, ball-iron-
stone, cornstone, afford examples of beds
naturally interrupted on the small scale.
Greenstone and killas generally form huge

" in America," &c. — Woodward's Essay towards a
Natural History of the Earth. Lond. 1702.

" This confirms what you say of the regular disposi-
" tion of the earth into like strata or layers of matter
" commonly through vast tracts." — Holloway's Letter
to Woodward in Phil. Tr. vol. xxxii. A. D. 1723.

insulated lumps in the midst of slate. Rock-salt occurs only in patches; the ochre beds of Shotover cannot be traced to any distance; nor the fullers earth of Rygate and Woburn; nor the Bath stone; nor the Hedington stone; nor any other with which we are acquainted.

The beds of coral rag, and calcareous grit, which form the southern bank of the Isis from Faringdon to Oxford, are lost a little to the north of that town, and do not shew themselves again on this side of Yorkshire.

At Ingleton; in the vale of Llangollen; at Ravenstone; at Plymouth, mountain limestone reposes immediately upon slate, to the exclusion of the old sandstone which appears on the south of Shrewsbury, at Sedberg, &c., and forms a considerable range of hills along the banks of the Wye and the Uske. This limestone is covered by coal measures in the northern counties; by lias in the neighbourhood of Frome; by the inferior oolite near Doulting; by the superior at Mells. At Chewton Mendip, oolite is wanting, and lias covered by forest marble. At Chard forest marble is also wanting, and lias covered by chalk. Near Charmouth, the green sand or mulatto rests on lias; at Blackdown on marl containing gypsum; the Purbeck and Portland beds are absent between Oxford and Abingdon, and the iron sand rests on Kimmeridge clay; the Kimmeridge clay is absent at Cumnor and Faringdon, and the iron sand rests upon the oolite of Hedington.

Greywacke and greywacke-slate are not seen, where they might be expected, along the northern edge of the secondary country which crosses Scotland from the firth of Forth to the mouth of the Clyde. In the county of Tyrone, the intermediate strata vanish in succession till chalk is very nearly in contact with granite.

Near Dresden, the primitive rocks are immediately covered either by coal-measures or by the quader-sandstein, a rock probably contemporaneous with chalk. Not far from Angers, gneiss and hornblend-rock are capped by a calcareous brecchia, abounding in shells and corals, apparently of the same date as the crag of Norfolk and Suffolk. In Auvergne, the primitive rocks are covered partly by coal, partly by a fresh water formation.

At Altenstein in Thuringia, bituminous marl slate, with impressions of fish, rests immediately on granite.

At Quebrada-sicca in South America, and in the Isle of Trinidad, Humboldt found mountain limestone lying on mica-slate: the former of these is almost every where deficient in France, and the latter in England.

It is a mistake, therefore, to suppose, that broken stratification is to be found only in porphyry, and the more recent varieties of trap ; and it is not a little extraordinary that the mistake should have originated in that school, which invented the distinction (with what propriety will soon appear) between partial and universal, principal and subordinate formations.

In Derbyshire, the limestone is so thick that it has never been sunk through ; in Yorkshire its thickness does not exceed twelve fathoms ; further northward, it is still thinner ; but a compensation is said to take. place by the intervention of sandstone and other beds which are wanting in Derbyshire. The northern part of the Holywell and Oswestry ridge consists of simple limestone; the southern, of limestone alternating with sandstone.

At Dudley the principal coal is ten yards in thickness; at a little distance they find, in lieu of it, three or four thinner seams of coal, with beds of sandstone interposed.

Enough has been said to make it evident, that neither any single stratum, nor single rock, nor any imaginable series of rocks can be traced in a continuous line round the globe. Similar strata, similar rocks, similar series of rocks are, however, found in different countries and in different hemispheres.

But will this similarity of character entitle us to suppose that they were once connected? products of the same æra? precipitates or deposits from the same solvent? Certainly not; for similar rocks are continually seen in very different formations. How often do we observe, in a mountainous district, recurring strata composed of the same substance, separated by a vast thickness of strata composed of other substances ! Is it not ascertained that the limestone of Melmerby Scar is more ancient than that of Alston ? that the red sandstone of Cheshire is less ancient than that on the banks of the Uske? that the green sand of Blackdown lies lower in the series than that of Feversham ? the oolite of Bath than that of the Isle of Portland ? In mineralogical character these rocks agree with each other ; and yet a mere agreement of mineralogical character has been thought sufficient to establish the identity of rocks situate at the opposite extremities of the globe !

The Shells which occur within the Basin of Paris, are said to occur also in Carolina and Virginia ; be it so ; are we to infer that the same Shell-bank once extended uninterruptedly across the Atlantic ?

It is probable that rocks deposited in places at no great distance from each other, at the same time, were not always of the same kind. There seems no reason, for instance, why the granite of Cornwall should be contemporaneous with the granite of the Pyrenees, rather than with the slate. In cases

where two rocks, commonly supposed to be
long to very different æras, are brought together, the series being incomplete, the insensible gradation, which these rocks display,
clearly evinces that there has been no pause,
no interval of time between their respective
births. A little north of Rother Bridge, in
Westmoreland, as has already been observed,
there is an intermixture of character in the
slate and limestone; although in the immediate neighbourhood these rocks are separated by the old red sandstone, and at
Ingleton, not very distant, they lie conformably to each other, and rounded pebbles of
the lower beds are enveloped in the upper.
It would seem, therefore, that the mountain limestone at Ingleton was deposited
at the same period as the old red, and not
at the same period as the mountain limestone near Rother Bridge; in other words,
that two beds agreeing in external character, containing the same fossils, and found
in the same neighbourhood, do not belong
to the same formation; while two beds having no such similarity in character or in
their fossil contents, do. At Argenton, in

France, lias passes in like manner into green sand or mulatto, and even partakes of its fossils. Either this green sand then must be coeval with our inferior oolite, or the lias not coeval with our lias. If we assume that the beds of mountain limestone in Derbyshire are the same as in Cumber·land, the toadstones of the one county must have been deposited at the same period as the hazels and plates of the other.

Unable to connect similar rocks of distant countries, obliged to connect dissimilar ones in the same neighbourhood, can any one uphold the doctrine of Universal Formations? Let him, who answers in the affirmative, reflect on the consequences which that doctrine involves. He must admit that, when the particles of quarz, feldspar, and mica, which had heretofore arranged themselves so as to form granite, changed their mode of arrangement so as to form gneiss, that change was conveyed with the rapidity of an electric shock from one end of the world to the other ; — that the currents

Q

of different hemispheres had so equable a
motion; that the particles borne along by
these currents were so equally assorted,
that, within the tropics, and without, the
same depositions began and ceased at the
same moment; — that similar pebbles were
detached from their native rocks, at the
poles and at the equator, by equal forces acting under the same circumstances, and were
deposited and cemented by the same means,
and at the same time. All this he must
admit, or reject *in toto* the doctrine of
Universal Formations.

It has been supposed, that the ᵃ analogy
observed in the rocks of different parts of
the world does not extend to the secondary;
but this opinion is erroneous. Coal occurs
in China and the East Indies; the Gypsum
of America agrees with that of Europe;
the Portland bed has been recognized in
the neighbourhood of Moscow; the Chalk
and Mulatto of Cracow correspond to the
Irish, and the Marlstone, which contains

Cuvier Disc. Prelim.

ammonites in Hindostan is undistinguishable from that of Lyme Regis, or Whitby.

In the scanty[a] catalogue of rocks with
which the Wernerians have furnished us, we
find some, as granite, which are common to
all climates ; some, as primitive gypsum and
serpentine, which are confined to a few spots;
some, as topaz rock and whitestone, which
are peculiar, or nearly so, to the neighbourhood of Freyberg. Yet we are told that the
primitive, transition, and flötz rocks are almost all Universal Formations.

I am not aware that any enumeration has
been published of the Formations which by
Werner were considered partial. The only
one adverted to by Mr. Jameson[b] is that
of Wehrau, in Lusatia, which consists of
sandstone, limestone, bituminous shale and
iron clay, resting on loose sand. I have
seen specimens of these in the collections

[a] So scanty that it does not contain even the fresh
water beds of Saxony.
[b] Jameson's Geology, p. 63.

at Freyberg and elsewhere. Similar substances with similar fossils are found
in this country ; but, whether in the same
relative position as at Wehrau, remains to
be ascertained.

" Porphyry, Sienite, and Basalt, (says Mr.
" Bakewell [a],) are evidently Partial Forma-
" tions, and have been produced by local
" causes, whose operations have been confined to particular districts ; no fact in
" Geology appears more decidedly esta-
" blished than this."

I shall probably have occasion to show
in the course of these Essays, that, so far
from being partial, the rocks, just cited, are
among those which are most extensively
diffused.

All the valuable coal-beds in Europe
have been found associated with a particular
series of rocks ; but there are many other
series, which contain coal of inferior quality, and in smaller proportion. In the one

---

[a] Bakewell's Geology, p. 119.

series the coal may be said to be principal, in the others accessory or subordinate. The unfortunate substitution of the term Independant for Principal has tended to bring into disrepute a division which of itself is useful. It may be right, however, to mention, that the words Principal and Subordinate must be here understood to refer only to the proportionate quantity and quality of coal in different formations, and not to the proportion of the coal as compared with other beds of the same formation. Even the ten-yard coal of Dudley, the thickest known, is very insignificant in comparison with the beds of sandstone and plate, with which it is associated.

The terms Principal and Subordinate, employed in the latter sense, may be useful when applied to small districts, but, applied to large, are at least premature. Minerals, which are extremely scarce in one country, are in another extremely abundant. Gneiss and mica-slate may properly be called subordinate in England, not so in Saxony. Primitive limestone may be called subordinate in Sweden, not so in Greece. Gypsum may be called subordinate in this country, not so in the South of Europe. Mountain limestone, the principal formation in Derbyshire, is in Herefordshire only a subordinate formation.

ESSAY V.

ON THE ORDER OF SUCCESSION IN ROCKS.

WE have seen that Formations are not universal, and that rocks found in different parts of the world, though similar, may be of different æras. We now proceed to a question not less important in a speculative, and far more important in a practical view. Let it be supposed, that certain rocks are known to occur in a certain district; will analogy enable us to predict the order of their occurrence? Do the rocks, of different countries, which resemble each other in external character, resemble each other also in relative position? or may a substance, which is superior to an adjacent substance at one place, be inferior to it at another ?

The question is not difficult. Every one admits that rocks alternate ; if so, they do not follow one another in regular order.

But. though every rock alternates with some others, it does not alternate with all. Flint alternates with chalk, clay with oolite, red marl with gypsum ; but no one, I presume, has seen granite alternating with salt, or serpentine with lias.

On the other hand, there is often such an affinity between two substances, that, on meeting with the one, we may speculate with a high degree of probability on the near occurrence of the other ; in this manner chert is associated with limestone ; rock-salt with gypsum ; coal with plate and gritstone.

Here, therefore, as in every other part of nature, we find uniformity and variety blended together ; the succession of strata is inconstant, but there is a limit to the inconstancy.

Lehman appears to have been the first

who attempted a chronological arrangement of rocks ; he divided them into two great classes — the Primitive and the Secondary.

To these, Werner added a third class, as partaking of the characters of both : Was this an improvement ? No : unless it would be an improvement to increase the list of primitive colours by the addition of mixed tints, or the list of notes in music by telling in the flats and the sharps. The object of classification is not to enfeeble distinctions, but to strengthen them. Hard lines are indispensable in science, the business of which is not to imitate nature, but make it understood.

The names which Werner assigned to his classes, are Primitive, Transition, and Flötz, or, to do ampler justice to the author's ideas by a more correct translation of his terms, Original[a], Intermediate, and Level.

The rocks of the first class, though all Original, are not all contemporaneous.

[a] Werner's Phraseology is unfortunate. Can there be a medium between primogeniture and horizontality ?

Granite is older than gneiss, gneiss than mica-slate, mica-slate than clay-slate, clay-slate than primitive trap. Of [a] Original limestone there are several formations. A deluge is said to have taken place between the birth of these rocks, and that of the superincumbent sienite and porphyry; yet the porphyry, sienite, limestone, trap, clay-slate, mica-slate, gneiss and granite, are all equally Original.

Upon these rest the Transition rocks, and upon them the Flötz, to each of which is assigned a definite place.

Virgil's peasant fondly imagined his native village a miniature of imperial Rome. With a corresponding love of generalization, Werner imagined Saxony and Bohemia a miniature of the world. He set down, more or less correctly, the order in which different rocks had arranged themselves in the districts with which he was acquainted, and concluded, that such must be the order of their arrangement in every other district. His theory was useful as a standard of reference, an incitement

[a] Jameson's Geognosy, p. 127.

to inquiry, a clue to observation; but, unfortunately for Werner, his pupils viewed it in a different light; it was represented by them not as an hypothesis to be tried, but as a system to be [a] followed; a system mature at its first conception, and perfect in all its parts and proportions. To merit of so high an order Werner was not entitled.

It was long supposed that the prop and stay of every other rock in every part of the world, must be Granite. The genius of Buffon led him to dispute this proposition. From the texture of granite he inferred that quarz must have existed previously,

[a] Heureusement pour les sciences la nature produit de tems en tems des grands hommes qui avec un courage égal au grand savoir, s'élèvent au dessus des ideés de leur siecle, et tracent le chemin pour les tems à venir. C'est ainsi que du cahors des masses éparses et pour la plupart inconnues, Werner a su batir le grand édifice de la Géognosie au quel il a attaché son nom.
Journal des Mines, t. 26. p. 168.
C'est dont un travail pour les siècles à venir, un triomphe pour quiconque saura aggrandir cette masse de connoissances, et un monument de gloire pour Werner, qui durera tant qu'il y aura une roche qui puisse attester le grand mérite de cette homme illustre.
Journal des Mines, t. 26. p. 198.

and that the disintegration of the quarz supplied a principal part of the materials of which granite was afterwards composed. Hacquet's[a] doubts upon the subject were better founded, though indistinctly expressed. Link[b] observed, that Granite, though it generally lies beneath other rocks, may not do so every where, and may, perhaps, in some places, even lie above them. It assumes too many forms to be every where contemporaneous. Granite will sometimes resemble granite less than it resembles porphyry or gritstone. When we consider at how late a period quarz has been formed in metallic lodes, we must not insist very strenuously (he says) on the extreme antiquity of granite.

Dolomieu, observing that the granite of Auvergne is covered immediately by lava, enquires, Whence did this lava originate? It is not probable that it originated in the granite; and, if not in the granite, there

[a] Reise, vol. ii. p. 46. 76.
[b] Anleitung zur geol. Kenntniss der Mineralien.

must be another rock still lower than the granite.

Saussure [a] remarks, that, although it may be true, as a general position, that granite is older than gneiss, it is evident from their mutual transition, inter-stratification, and conformity of dip, that these two rocks are occasionally coeval.

The alternation of granite and gneiss has been since observed in various countries; at Naundorf [b] near Freyberg ; at the Schneekoppe [c] in the Riesengebirge; at Karpenstein [d] in Bohemia; in the Alps; in South America. [e]

Fragments of gneiss are found in granite at the Brocken, the most consider-

[a] Voyages, tom. ii. p. 228.
[b] By Karsten, Ubersetzung von Peyrouse, p. 20. note.
[c] Jameson in Nicholson's Journal, vol. ii. p. 228. Charpentier's Beitrag zur geognostischen Kenntniss des Riesengebirges, part iv. p. 54. 63. Meuder's min. Reise und Einleitung. — Gerhard's Abhandlung uber die Umwandelung einer Erd und Steinart in die andere, p. 109.
[d] Von Buch's geogn. Beobachtungen, vol. i. p. 23. and Landeck; see Anderson's Observations in his translation, p. 23 and 113.
[e] Humboldt's Geographie des Plantes, p. 123.

able of the Harz[a] mountains, and in the neighbourhood of Vienna.

Granite alternates with gneiss and mica slate on the south[b] of the Taberg; [c]with eurite, or white stone in the Saxon Erzgebirge.

It rests on mica-slate at Reichenstein[d], in the county of Glatz ; at the foot of the Eulengebirge ; along the entire ridge of the Riesengebirge; in Moravia and the Tyrol.

It alternates with mica-slate at Lugnaguilla in the county of Wicklow, and at Derryclare in Conemara.

That granite frequently alternates with Shist, is attested by Pallas[e], Soulavie[f], La Peyrouse[g], and Saussure.[h]

[a] Schubert's Geognosie, p. 110.
[b] Napione in Bergm. Journ. vol. ii. p. 2006.
[c] Raumer's geogn. Fragmente.
[d] Von Buch's Landeck translation, p. 44 and note. Geogn. Beob. vol. i. p. 37.   Schubert's Geogn. p. 110.
[e] Helvet. Mag. 175. 2 Pallas Reise, 517. 520.
[f] Soulavie, vol. iii. p. 162.
[g] La Peyrouse sur les Mines de Fer. 325.
[h] Voyages, § 662 and § 676.

According to Jameson[a], the granite mountains in Arran rest on clay-slate.

A rock, differing from granite only in the imperfect cohesion of its ingredients, is found, in many parts of Cornwall and Devonshire, incumbent upon killas. Alternations of the two substances are seen at Huel Fortune near Marazion, and at the more celebrated mines of Cooks-kitchen and Dolcoath. At Gwarnock, in the parish of St. Allan, a steatitic granite accompanies the lode, which is worked in killas. The Morwelham tunnel, driven through a killas country, traverses three distinct beds of grauany elvan. At Cliggar point, on the east of St. Agnes, a rock of the same nature lies upon killas.

At a small cove, called La Poulet, in the neighbourhood of Cherbourg, I have observed killas passing into granite, and dipping beneath it. Of the intermixture of these substances numerous examples may be found in Cornwall; in Wicklow; in the

[a] Von Buch's Norway, Black's translation, p. 139, note.

5

Mourne mountains; at the Lowren-hill in
Galloway; in the Grampians; in the Isle
of Arran, &c.

Subordinate beds of hornblend-rock occur
in granite, at Gwindu in Anglesea; and at
Kilranelagh in the county of Wicklow.

Granite rests upon hornstone slate
(hornschiefer) at Ehrenberg[a]; it alternates
with it near Tschito[b] on the banks of the
river Ingoda; and at the Chalet de la Para
near Chamony[c]. At Kandy[d], a village in
Tartary, it alternates with hornstone; at
St. Michael's Mount in Cornwall with
quarz; in the Pyrenees[e] it rests upon serpentine.

Granite reposes upon limestone in various parts of France[f], in Corsica[g], in the

Voigt's min. und bergman. Abhandl. vol. i. p. 144.
[b] Patrin. Journal de Physique, vol. xxxviii. p. 229
and 289.
[c] Saussure's Voyages, § 676.
[d] Patrin. Journal de Physique, vol. xxxviii. p. 229,
and p. 289.
[e] Peyrouse sur les Mines de Fer. p. 329.
[f] Bertrand Journal des Mines, vol. vii. p. 376.—He
appeals to Faujas and Soulavie.
[g] Dolomieu Journal de Physique, vol. xxxix. p. 9.

Tyrol [a], and still more frequently in the Pyrenees.[b] Soulavie refers to an adit driven from one of these rocks into the other, so that there can be no mistake as to their relative position. A later writer tells us that in the ridge last mentioned, granite, porphyry, trap, corneenne, petrosilex alternate with limestone, and are so intermixed with it at the point of contact, that it is impossible not to consider all these substances contemporaneous. At one place a bed of granite, from 20 to 25 centimetres in thickness, is incased in a bed of trap, which is itself incased in limestone. The trap dwindles away till the included granite touches the limestone, which is then penetrated by zigzag veins and nodules of granite; these two rocks at their junction are firmly cemented together.

It is said that between Weisbaden and Idstein, M. Habel found a fossil shell in

[a] Ann. de Chym. vol. xiii. p. 166. Duhamel, Journ. des Mines, vol. viii. p. 751. and 756. See also Palassou's Description des Pyreneés.

[b] Collini, Considerations sur les Monts Volcaniques. Bergman's Manuel, vol. ii. p. 272.

R

granite : Is the specimen now in existence?

" I found in the neighbourhood of Chris-
" tiania, says M. von Buch [a], rocks evidently
" belonging to the transition series, which
" till now no one ever suspected to belong
" to it : immense mountains of porphyry
" resting upon a limestone, full of fossils ;
" over this porphyry, sienite and granite,
" in composition agreeing entirely with
" that of the oldest formation. Granite
" above transition limestone ! Granite a
" member of the transition series !"

Messrs. Brongniart[b] and Omalius d'Halloy
consider the granites in the Cotentin more
modern than the adjacent slate and limestone, in which are found organic remains.

On the eastern[c] side of the Erzgebirge,
granite rests on greywacke and greywacke
slate ; it probably occupies the same position in Somersetshire, Caernarvonshire, and
Worcestershire.

[a] Von Buch's Travels in Norway. — transl. p. 45.
[b] Journal des Mines, tom. xxxv. p. 126.
[c] Raumer's geogn. Fragmente, Bonnard in Journal des Mines. tom. 38. Ann. de Chemie, vol. i. p. 210.

" In the Saxon Erzgebirge[a], says Jame-
" son, we observe the oldest gneiss covered
" by clay-slate, which contains beds of flinty
" slate, greenstone, limestone, and por-
" phyry; over these rests, in a conformable
" position, newer granite, which alternates
" with sienite, gneiss, and porphyry."

Analogous observations have been made,
in the Alps, by Ebel, Escher, and Saussure;
in the Thuringian Forest, by Heim; in Scotland and its Islands, by Professor Jameson,
and Dr. MacCulloch.

I had occasion some time since to visit Les
Trois Couronnes, situated on the north eastern frontier of Spain. This mountain is
well known: Mina was encamped upon it
during the late war, and it has been, in
earlier periods, repeatedly the scene of important events, both military and political:
from the boldness of its out-line, and its
proximity to the sea, this mountain forms
one of the most conspicuous objects on the
eastern side of the Pyrenees.

Ascending from the ferry over the Bida-

[a] Von Buch's Travels in Norway, p. 283. *Note.*

soa, we first met with a rock of mountain limestone, and afterwards with a conglomerate, made up of pebbles and grains of very various sizes, all derived apparently from the debris of older rocks in the neighbourhood. It contains pieces of black slate, and alternates with thin beds of this substance. Higher up, the beds of sandstone or conglomerate become less frequent, till at last this rock entirely ceases, and the slate, which lately alternated with it, becomes the prevailing rock of the district. The different beds of this assume different appearances; some, which have the aspect of greywacke-slate, from the broken scales of mica, irregularly disseminated through their substance, are interposed among others, which, from their glossy and uniform surface, seem to have a claim no less decided to a place among the clay-slates; some might fairly rank with mica-slate; and the older members of this series alternate with granite, which, at first occurring in diminutive beds, gradually acquires importance, and at last constitutes the three summits, which give name to the mountain.

This granite, wherever I had an opportunity of examining it, exhibited the same characters; small grained, imperfectly crystalline, prone to decomposition, of a yellowish grey tint, it very much resembled the common grauan of Cornwall, to which it is also analogous in its metallic lodes, tin and copper having formerly been worked in it.[a] Its alternations with the slate or killas are numerous and distinct; in some places the beds of granite and slate do not exceed a yard in thickness; but they are evidently beds, not veins, for they lie parallel to each other.

The Wernerians acknowledge three formations of granite. The first is gratuitously supposed to be fundamental. I say gratuitously, for, by the terms of the proposition, the bottom of this formation has never been seen, and consequently we have no means of ascertaining whether it be fundamental or not.

[a] In addition to the substances already noticed, there are upon this mountain an elvan-greenstone and hœmatitic iron-stone.

The second consists of veins traversing the former. Are these veins, which evidently bear to granite the same relation as calcareous veins to limestone, entitled to the dignity of being called a formation ?

" The third occurs sometimes· in the " state of veins, sometimes of uncomform- " able beds :" if so, might it not be properly divided into at least two formations? and might not one of these be allowed to merge in that last mentioned ?

These three formations appear to me useless. It is more simple to say, the relative position of granite is inconstant.

That equal uncertainty prevails in the relative position of the other rocks usually denominated primitive, will be rendered evident, I think, by the following sections.

Section from the Junction of the Isere and Ar, in Savoy, along the Valley of Maurienne, over Mont Cenis, to Avigliano.

On the North of Mont Cenis.

Mica-slate
Gneiss

Compact Feldspar
Siliceous Slate, or lamellar feldspar
Primitive Gypsum
Primitive Clay-slate
Mica-slate
Primitive limestone
Primitive clay-slate
Mica-slate
Primitive limestone and gypsum
Gneiss
Primitive limestone
Mica-slate
Primitive Gypsum

Section on the South Side of Mont Cenis.

Mica-slate	Alternating all
Primitive limestone	the way from the
Serpentine	summit of Mont Cenis to Nova-
Primitive clay-slate	lese.

Quartz

Primitive limestone	
Serpentine	Alternating.
Mica-slate	

Veined Granite

Serpentine and other varieties of magnesian rocks extending to Avigliano.

Section *from St. Maurice, in Unterwald, through the*
Valley of Antremont, over the Great St. Bernard,
through the Vale of Aosta to Yvrea.

On the North of the Great St. Bernard.

Gneiss
Lamellar feldspar
Compact feldspar
Coarse-grained Greywacke
Fine-grained Greywacke
Primitive Clay-slate
Coarse-grained Greywacke
Mica Slate
Gneiss
Lamellar feldspar
Primitive limestone
Mica-slate
Gneiss
Primitive Gypsum
Primitive Clay-slate
Primitive Limestone ⎫
Gneiss   ⎬ Alternating.
Mica-slate   ⎭
Gneiss
Hornblend Slate
Primitive Clay-slate

White and black Quartz
Mica-slate with Garnets.

———

On the South of the Great St. Bernard.

Primitive Clay-slate with veins of
    Gypsum
Mica-slate
Gneiss
Quartz
Primitive Limestone
Gneiss
Primitive Limestone
Hornstone
Mica-slate
Potstone
Actinolite, with or without ⎫
    Garnets   ⎬ Alternating.
Primitive Limestone   ⎭
Gneiss
Mica-slate
Primitive Limestone
Primitive Greenstone and Serpentine.

In Ebel's work, from which these sections have been taken, will be found several

other sections, tending to the same conclusion.

———

So much in regard to the Alps; proceed
we now to the Pyrenees.

" The Pic du Midi de Bigorre[a] consists
" entirely of primitive rocks in continuous
" distinct beds, dipping between 60° and
" 80° from the general chain of the Py-
" renees.

" The lower beds consist of limestone,
" alternating several times with compact
" feldspar (roche de corne) and perhaps
" trap.

" The upper beds are micaceous gneiss,
" and grenatite.

" Above the gneiss are a great many
" alternate beds of limestone, trap, roche
" de corne, and now and then of granite.

" The roches de corne often assume the
" most fanciful curves, though lying be-
" tween beds of limestone, whose strata
" are plain.

" Granite occurs in the upper beds, as
" a vein, as a bed, and as an ingredient in

[a] Duhamel, Journal des Mines, vol. viii.

" the limestones: in this last case it is
" found only on the surface, as if deposited
" immediately after the consolidation of
" the calcareous molecules." [a]

---

*The following is the Order which the Primitive Rocks
observe in the Saxon Erzgebirge.*

Granite [b]
Gneiss
Mica-slate
Talc and Alum-slate
Clay-slate, with beds of Greywacke-
    slate, corneous Trap, Greenstone,
    and Lydianstone
Greywacke, with beds of Gneiss, Por-
    phyry, Granite, Limestone, Slate, or
    Granite
Slate
Granite passing into Sienite, with
    subordinate beds of Gneiss, or of
    Porphyry and Limestone.

In various parts of Saxony you find

[a] Schubert, p. 104.
[b] Werner Kuzze Classification 14 — Charpentier 55.
57. 174. 201. 400.

Limestone lying under Gneiss, to the depth of 200 feet, or alternating with Mica-slate and Clay-slate.

At Klostergrab[a], in Bohemia, Gneiss rests upon Porphyry.

Beds of Hornblend, under different varieties, says Von Buch, appear to be subordinate to all the primitive formations.

If we consult Herman on the Urals, Von Buch and Hausman on the Norwegian Chain; Brongniart on the Cotentin; Raumer on the Harz; Heim on the Forest of Thuringia; Fichtel on the Carpathians; Dolomieu on the Vosges; we shall find, in each of these districts, anomalies no less remarkable than those which have been already detailed. In America, the order of succession, which the Wernerian theory prescribes to the primitive rocks, is so often varied and reversed, that Maclure[b] declares it impossible to arrange them in any regular series.

Groscke[c] states, that the Grampians pre-

[a] Voigt.
[b] Journal de Physique, vol. lxxii. p. 142.
[c] Bergbaukunde, vol. i. § 399.

sent alternate rocks of Gneiss, Clay-slate, Mica-slate, Clay-slate, Porphyry, and Granite; and the more detailed and accurate observations of Dr. MacCulloch and Professor Jameson have sufficiently established the uncertainty, which prevails, in the succession of the different primitive rocks, in the northern portion of our island.

According to the principle of arrangement adopted at the Geological Society, the several specimens ought to observe the same order in the cabinet, which they observe in nature; no attempt, however, has been made there, as yet, to arrange any of the primitive substances upon that principle; it is reasonable, therefore, to infer, that, in England, the order in which these substances occur, is either uncertain or unknown.

In Ireland, the order of succession of these substances is extremely variable; the Wicklow mountains are very differently constructed from those of Downshire, and the Downshire from those of Donegal, and Tyrone.

The primitive rocks ought to be followed by the transition, but almost every country, that has been well examined, presents numerous instances of transition rocks occurring in the midst of primitive districts, or primitive in the midst of transition. The writings of Brochant, Brongniart, Raumer, Von Buch, &c. abound in instances of this kind. The slate of Snowdon, and that of Tintagel contain organic remains.

It is commonly thought by foreign writers, that the secondary beds are more capricious, in the order of their occurrence, than the primitive and transition. A contrary opinion prevails in England. It is here supposed, that the secondary beds are extremely regular. Their irregularity however is very problematical.

The various Clays and Marls, which alternate with coal, lias, the different oolites, mulatto and chalk, are scarcely distinguishable from one another in their aspect or their properties.

Striated Gypsum is represented by the Freyberg school, as characteristic of a particular formation; yet, by some of the pupils of that school, it is placed beneath, by others, above the second flötz limestone. In England it is found associated not only with red marl, but with the Purbeck beds and the London clay.

Green sand or mulatto lies beneath the iron sand at Brill, and Apsley Guise; at Pusey, above the iron sand, but beneath the chalk; at Reading above the chalk, but beneath the London Clay, at Bagshot-heath above them all.

If Trap had a definite position in the series, it would not be found in the old sandstone of Herefordshire, in the mountain limestone of Derbyshire, in the coal measures of Staffordshire, in the lias of Perthshire, in the chalk of Kinbain.

At the Clee hills, and on the banks of the Avon near Bristol, the Mountain Limestone is oolitic; the Lias is oolitic in Glamorganshire; at Hambden Hill in Somersetshire, the Bastard Freestone is oolitic. The Bath stone is an oolite; the Heddington stone an oolite; the Portland stone an oolite; has the Oolite then any determinate place in the order of succession?

In Saxony and Silesia Coal is said, and I believe with reason, to lie beneath the old red rock ; in the North of England it lies sometimes above the mountain limestone, sometimes beneath it. That, which is raised in the eastern moorlands of Yorkshire, is associated with rocks of a more recent epoch even than the lias. There is scarcely one of the secondary beds in which coal has not been observed, although perhaps of bad quality, and in beds of insufficient thickness to pay the expence of working them.

Even between the primary and secondary " formations" the line of demarcation is far less distinct than has generally been supposed. Green-stone and slate are found in every formation. I have a specimen from the West Indies, of a greenstone, which has every character of the primitive except that it contains shells. Hornstone porphyry occurs, as a flötz rock, in Arran ; granular marble, in Dalmatia; serpentine, in Fifeshire ; mica-slate, in the Valais ; granite and sienite, in Saxony and Norway.[a] On

---

[a] Wild's Salines, p. 75.

the other hand, quartz-rock occurs as frequently in the coal formation, or in beds which cover the chalk of the basin of Paris, as it does in the primitive districts of Schichallion, Wicklow, and Conemara.

It is said in the Wernerian theory, that, after the formation of all other strata, an immense deluge suddenly occurred, and as suddenly retired, leaving, behind it, those scattered hummocks of flötz-trap, which have, for some years, so greatly engaged the attention of geologists.

The proofs of this catastrophe, we are informed, are to be found in the great elevation which these rocks occasionally attain ; in their broken stratification ; in their unconformable posture ; and in the nature of their materials.

But are trap-rocks really more elevated than others ? or their stratification more broken ? It is time enough to consider inferences when we have established facts.

If the posture of trap is often unconformable, so is that of granite, sienite, horn-

s

blend rock, porphyry, primitive green-stone, &c.

Every rock without exception lies, some-times, in a conformable, sometimes, in an unconformable posture : and perhaps the different members of the flötz-trap form-ation, as often exhibit a want of conformity towards each other, as, towards the beds on which they repose.

But the nature of its materials. — Many of them are precisely the same, as those found in other formations. The only rocks, which are cited as peculiar to, and charac-teristic of, the newest flötz-trap, are basalt, wacke, greystone, porphyry-slate, and trap-tuff. I am not sure, that I know what grey-stone is ; the only locality, given of it by Jameson, is Vesuvius, where it is said to form a portion of the unchanged rocks. The doctrine that it belongs to the flötz-trap, therefore, is founded on an assump-tion, that we have the means of distinguish-ing, in volcanic countries, substances, which have been changed by the volcano, from those which have not : an assumption somewhat gratuitous. The remaining substances, viz. basalt, wacke, porphyry-slate, and trap-tuff, are certainly not peculiar to this formation, as in England, Scotland, and Ireland, they are, often, found interstratified with other formations much older. There is reason to suspect that, in Germany, trap-rocks, of very different æras, have been referred to the same æra, and that much of that which has been supposed the newest flötz-trap in Scotland, and which ought, therefore, to be more modern than the beds of the basin of Paris, is coeval with red sandstone, mountainlime-stone, and coal.

ESSAY VI.

ON THE PROPERTIES OF ROCKS, AS CON-NECTED WITH THEIR RESPECTIVE AGES.

I. ON THE INGREDIENTS OF ROCKS.

It was a favourite idea some time since, that the simple earths [a] were older than the compound.

It was supposed, also, that siliceous [b] rocks were older than calcareous. M. Lefebre [c] and M. Dolomieu took great pains in 1791, to establish the then-disputed existence of primitive lime-stone.

These doctrines it is unnecessary to dis-

[a] Journal de Physique, vol. xxxix. p. 374.
[b] L'Encyl. Geographie Physique, vol. i. p. 96.
[c] Journal de Physique, vol. xxxix.

cuss now, as they have no longer any adherents.

As far as our present experience reaches, granite and gneiss seem to belong, peculiarly, though not exclusively, to the more ancient rocks : chalk, clay, sand, marl,* loam, rock-salt, to the more modern. Greywacke, sandstone, clay-slate, quartz-rock, sienite, porphyry, greenstone, basalt, serpentine, compact feldspar, seem common to both. In general, the younger rocks exhibit more abraded fragments than the others, more bituminous and saline matter, more organic remains.

2. ON THE STRUCTURE OF ROCKS.

It has been thought that the primitive rocks were formed by chemical, the flötz by mechanical action ; and that, in transition rocks, the two actions were combined.

Of these three opinions, the last, though attacked in the Edinburgh ª Review, is

ª Ed. Review, Vol. ii. p. 343.

perhaps the only one, which is well founded.

La Metherie [a] observes, that, during the crystallization of primitive rocks, the earths were not all dissolved, and could not all crystallize; for in these rocks we often find a quantity of argillaceous matter in an earthy state. Werner speaks of gneiss containing fragments of granite; and Saussure has shown that brecchias prevail in rocks of every age. Mr. Playfair [b] quotes several instances of arenaceous rocks inter-stratified with rocks decidedly primitive; and greywacke is now very generally admitted into the list of primitive formations.

It has been proposed to substitute for the terms primitive and secondary rocks, crystalline [c], and deposititious; but, unfortunately, rock salt is as crystalline as clay slate,

[1] Theorie de la Terre, vol. v. p. 25.
[b] Illustrations. See also on this subject Dr. MacCulloch's able account of quartz rock in Geolog. Trans. vol. ii. & iv.
[c] M'Lure, Journal de Physique, vol. xii. p. 145.

sulphur as porphyry, the Fontainbleau sandstone and burrstone as greywacke, gypsum as serpentine. Mechanical action seems to have commenced at a very early period, and chemical to have continued up to the formation of the latest strata.

Dolomieu[a] was, I believe, the first who made the conglomerate, or old red sand-stone, the line of demarcation between the primitive rocks and the secondary.

The structure of rocks changes, by almost imperceptible gradation, from granular to compact, from compact to porphyritic, from porphyritic to amygdaloidal, &c. In general, however, the crystalline rocks may be said to be the oldest; the sandy, marly, clayey, the newest.

The granular varieties of limestone are, for the most part, older than the compact, and the compact older than the earthy.

Slate occurs in all formations; but the more perfect varieties seem confined to the primitive and transition rocks.

[a] Journal de Physique, vol. xxxix. p. 390.

3. ON THE SPECIFIC GRAVITY OF ROCKS.

It was thought, during the infancy of geological science, that the order of succession in rocks was that of their specific gravities.[a] Even Woodward, with all his experience, fell into this error. Mr. Hawksbee exposed it at the Royal Society, by simply exhibiting the section of a coal mine.

Considered on the great scale, however, this hypothesis is not, perhaps, altogether erroneous. Generally speaking, the primitive rocks are of greater specific gravity than the secondary; and it appears from the experiments of Cavendish and Maskelyne, that the density of the superficial parts of the globe is less than that of its interior.

4. ON THE CONSOLIDATION OF ROCKS.

Nothing is more common, among the newer rocks, than the alternation of

[a] Varenius lib. i. cap. 7. propos. 7. Hawksbee's Experim. p. 317. Luidii Lythophil. p. 110.

consolidated beds with beds not consolidated.

On the other hand, all the older beds, or
nearly all, are consolidated. The fuller's
earth found at Rosswein in Upper Saxony,
and a similar bed found on the Old Man
mountain, at Coniston, occur to me as
exceptions to the rule ; but, in general, the
old red sandstone may be considered the
earliest formation, which appears unconsolidated.

As a reason for admitting the igneous
origin of trap rocks, it has been [a] said, that,
" of all other formations, the degree of
" consolidation decreases together with its
" age, their texture passing from crystal-
" line through the several gradations of sub-
" crystalline, compact, coarse, and lastly
" earthy, while in the trap formation, even
" where it rests on chalk, the crystalline
" texture of the oldest rocks frequently
" recurs."

I suspect that this distinction is not warranted by fact, and that trap rocks are, in

[a] Geological Transactions, vol. iii. p. 208.

regard to consolidation, analogous to all others ; at least, I am not aware, that the successive beds of any formation, which have fallen within the limited scope of my experience, present that regular increase of consolidation, which, with a single exception, is here attributed to all.

5. ON THE STRATIFICATION OF ROCKS.

The Huttonians distinguish carefully between stratified rocks and unstratified. The former, according to them, have been deposited by water, and merely hardened by Plutonic heat; whereas the latter have been thrown up in a melted state from beneath, and forcibly injected amid the pre-existing strata. It is a fatal objection to this hypothesis, that consolidated beds often alternate with unconsolidated, stratified with unstratified.

How far the opinion, which the Huttonians entertain, on the origin of the unstratified rocks, derives support from the phenomena of granitic or basaltic dykes, this is not the place to enquire. I must confine myself, at present, to an examination of arguments, which they deduce from the nature of the beds themselves. Now what are those arguments ? "that the " rocks in question are penetrated by pyrites " — that they contain fragments — that " the beds contiguous are sometimes bent, " disturbed or indurated — that the strata " above them are sometimes similar to " those beneath." Let us grant all this : does the conclusion follow from the premises ? or is there any one property, among those here attributed to the unstratified rocks, which may not, equally, be found among rocks that are stratified ?

On the other hand, the connection which subsists between beds of granite and gneiss, greenstone and compact feldspar, trap-tufa and clay, toadstone and cornstone, is too intimate to be considered only accidental.

At le Gros Cattel, near Cherbourg, a large-grained granite, consisting of flesh-coloured feldspar, greyish-blue quartz, and black mica, with some tourmalin, cor-

l

responding, in its general aspect, with the granite of Galway, is intersected, in different directions, by veins of milk-white or blueish quartz, having the gelatinous look of gypsum. These veins vary in breadth from an inch to a foot, and lose themselves, before they have run to any considerable distance. The granite continues, uninterruptedly, till you have passed the eastern promontory of a small cove, called le Poulet, where it is broken off, a few crags still appearing, however, about the middle of the cove. The easternmost of these is made up of the same ingredients as the granite abovementioned ; but the feldspar is paler, and the texture of the rock slaty. It appears to dip beneath the granite, and to rest upon killas. At the western promontory, the killas, like the granite, is traversed by numerous veins of blue quartz, which, in the more contorted beds, occurs, also, as a constituent part forming stripes.

I consider the prevalence of this blue gelatinous quartz, in the killas, as well as the granite, though separated from each

other by a bed of gneiss, a strong reason
for believing, that these three rocks were
formed at nearly the same epoch, and in
the same manner.

The dip of the killas is equally adverse
to the Huttonian hypothesis. If the granite had, naturally, no connection with its
adjacent rocks, but was thrown up, accidentally, in the midst of them, the strata
of killas should have dipped the contrary
way.

6. ON THE POSTURE OF ROCKS IN REGARD TO
THE HORIZON.

Vertical, and highly inclined rocks are
reputed older than those which are
horizontal.

" Next in the order of time to the con-
" solidation of the primary strata," says
Professor Playfair[a], " we must place their
" elevation, when, from being horizontal,
" and at the bottom of the sea, they were
" broken, set on edge, and raised to the
" surface : it is even probable, that to

Illustrations, p. 123.

" this succeeded a depression of the same
" strata, and a second elevation ; so that
" they have twice visited the superior, and
" twice, the inferior regions. During the
" second immersion, were formed, first, the
" great bodies of puddingstone, that, in so
" many instances, lie immediately above
" them, and, next,^{V} were deposited the
" strata, that are strictly denominated
" secondary."

M. Omalius d'Halloy [a] represents the
coal measures, and all the strata beneath,
to be inclined ; the red ground, and all the
strata above, to be horizontal. The inclined
beds of a mineral basin are, always, older,
he says, than those beds which are parallel
to the horizon.

I have shown, in a former essay, that
every species of rock assumes, occasionally,
every posture; hence, the inclination of
rocks forms no evidence as to their
antiquity. The observations, which I have
made, since that essay was written, confirm
me in the opinion I then entertained, that

[a] Journal des Mines, tom. xxiv. p. 154.

the extreme difficulty of distinguishing planes of stratification from planes of cleavage, in the slate rocks of inland districts, where small sections, only, are exposed to view, induce us, frequently, to suppose beds vertical, or highly inclined, which are, in point of fact, nearly horizontal.

Ramond [a] assures us, that it is rare to find, in the Pyrenees, at the base of the mountains, or at their summit, in the centre of the chain, or at its extremity, any secondary rock, whose strata form with the horizon an angle of less than 45°.

7. ON THE POSTURE OF ROCKS RELATIVELY TO ONE ANOTHER.

It is often, supposed, that unconformity of posture, in two adjoining beds, proves them to have been formed at different æras. This doctrine I consider erroneous for the following reasons :

1. Sienite, porphyry, basalt, all those rocks, for the decyphering of which so

[a] Journal des Mines, vol. xii. p. 88.

many deluges have been set in motion,
occur, indeed, sometimes, in an unconformable, but, often, in a conformable
position.

Of the conformable stratification of
sienite and porphyry, Professor Raumer
has given us a striking instance in the
neighbourhood of Freyberg. Heim, I
think, has noticed another in the forest of
Thuringia.

The basaltic rocks of Antrim are conformable to the beds, on which they
repose.

2. Of rocks the most intimately connected, many occur, almost always, in an
unconformable posture. Gneiss or mica
slate is, often, unconformable to granite;
greenstone slate to greenstone. What the
Wernerians call unconformable and overlying stratification, is, often, only the incur-
sion of unstratified or clotted masses, in
masses regularly disposed.

3. Different strata of the same bed are,
often, unconformable to each other. Wit-
ness those of red sandstone, of Pennant-
stone in the forest of Dean, of chalk at

Handfast point; or the laminæ unconformable to the strata, as at Swanwick,
Mells, Hedington near Oxford, Anthony-
hill near Bath.

The rock salt of Bocknia is said
to lie as represented in the following
diagram.

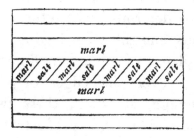

4. The same rocks will occur in different
parts of their course, both in a conformable,
and in an unconformable position.

Dr. MacCulloch has remarked that trap
will sometimes appear in the form of a
bed running parallel to the adjacent strata,
and then suddenly rise through them in
the form of a dyke. Similar phænomena,
though on a smaller scale, are not unusually

afforded by quartz, amid beds of primitive slate.

The transition and flötz rocks, which, according to a writer in the Transactions of the Royal Society of Edinburgh, are never conformable to each other, are, nevertheless, so strictly conformable along the frontier of Wales, at least to my eyes, that, after repeated examination, I am still unable to determine where the one formation ends, and the other begins.

The planes of mountain limestone, at Ingleton, cover the edges of greywacke slate, though, at Orton, these two rocks are conformable, and graduate into one another.

In some situations unconformity of posture, far from proving that the strata, in which it is observed, were formed at different æras, tends rather to prove that these strata were formed simultaneously. It is scarcely possible to imagine how the primitive limestone of Conemara can have been formed at a different period from that which gave birth to the rocks with which it is associated; though its position in regard to them is the most eccentric that can be conceived. I feel no less persuaded that the slate and quartz, at Ilfracombe, are contemporaneous, though their respective strata appear to meet nearly at right angles.

At Port Vieil[a], in the Pyrenees, the limestone makes its way through gritstone; and at Tuccaroy, vertical beds of this last rock are traversed by horizontal beds of limestone: yet, says the author to whom we are indebted for a description of that district, " it would be wrong to " suppose the one merely deposited on " the other. Both were deposited at the " same moment, and from the same pre- " cipitate : the two substances are blended " together along the line of junction, and " the intermixture continues to a certain " distance on each side of it."

In all these cases unconformity appears to have been occasioned by the disturbance

[a] Picot de la Peyrouse, Journal des Mines, vol. vii. p. 55.

T 2

which one of the substances sustained from the deposition or precipitation of the other.

8. ON THE DIP AND DIRECTION OF ROCKS.

The difficulty of determining these in primitive rocks has been already stated. Humboldt has hazarded an opinion that the primitive[a] rocks, both in Europe and America, run N.E. and S.W., forming with the meridian an angle of 50°, and dip to N.W., at an angle varying from 60° to 80°.

Dolomieu[b], however, did not find this to be the case in the Alps, nor Cordier[c], in the Pyrenees; and succeeding [d]travellers in other parts have multiplied exceptions to the rule till it has become useless.

The secondary rocks in the [e]Pyrenees

[a] Journal de Physique, vol. liii. p. 46.
[b] Ibid. vol. xlvi. p. 423.
[c] Journal des Mines, vol. xvi. p. 252 and 281.
[d] See Illustrations of the Huttonian Theory, p. 228.
[e] Journal des Mines, vol. xii. p. 88.

agree in direction with the primitive; the same conformity, if I am not mistaken, prevails also in England.

9. ON THE ALTITUDE OF ROCKS.

According to the Wernerian theory, a succession of rocks is always accompanied by a diminution of level, the waters from which they were deposited having subsided progressively during the period of their formation. The outgoings of the newer strata, therefore, are lower than the outgoings of the older, from granite downwards to the alluvial depositions; the trap rocks, however, being on this, as on every other occasion, a privileged order, not amenable to the laws imposed upon the rest of created matter.

With this one exception, however, we ought to be able to determine the relative age of rocks, by comparing their height.

La Metherie[a], who adopted this doctrine without sufficient examination, appears a

---

[a] Theorie de la Terre, vol. iv. p. 371.

T 3

good deal embarrassed in his attempt to sustain it. His assertion, that the granite of the Pyrenees is higher than the limestone, is followed by an admission that the limestone also is very high, sometimes higher than the granite. He accounts for this by assuming, that the granite is the more perishable of the two, and by supposing that it must have been the higher formerly, though the lower now.

The secondary ridge of the Jura is higher than the primitive ridges of the Vosges and Bretagny.

Mont Blanc [a], the highest mountain in Europe, is said not to consist of granite. Be this as it may, its summit is only 128 feet higher than that of Mont Rosa composed of stratified rocks.

The principal heights of the Apennines are of secondary limestone.

In the British islands the principal eminences, Ben Nevis, Snowdon, the rocks of Killarney, and the mountains of Cumberland, are composed either of porphyry

---

[a] Saussure's Voyages, § 2135. Playfair's Illustr. p. 200.

slate and compact feldspar, or of granular quartz, or of slate marked with fossil impressions.

The Ocrynian and Charnwood hills, composed of granite or sienite, are lower than the Pennine or central hills, composed of secondary limestone and coal measures.

MacClure affirms, that there is no rule, as to the height of the several primitive rocks in North America.

Near Huanco [a], in Peru, coal measures attain the height of 14,700 feet, the highest granite in that country being 11,500 feet high. At the Magdalen river, north of Quito [b], coal occurs at an elevation of 12,000 feet above the level of the sea.

Mount Ararat, Etna, the peak of Teneriffe, and the greatest eminences of the Andes, are volcanic.

La Metherie [c] is prodigal of reasons why gneiss should be found at great elevations,

[a] See Andes, in Supplement to Encycl. Brit.
[b] Journal de Physique, vol. xxxviii. p. 30.
[c] Theorie de la Terre, vol. v. p. 11.

as if not aware, that it is also found below
the level of the sea.

10. ON THE METALS CONTAINED IN ROCKS.

The German miners divide rocks into
flötz-gebirge, where the ore occurs in
beds, and gang-gebirge, where it occurs in
veins.

With the exception of pyrites, metallic
ores are rare in the newer rocks. Galena
and blend in small quantity have been
found, however, in the coral rag of Yorkshire; and galena, calamine, and iron ore,
are raised in Silesia from a rock more
modern than lias.

Tin, molybdæna, tungsten, wolfram, are
said to be confined to the oldest rocks.
Uranium and titanium are nearly contemporary with these.

Dolomieu denies that rocks can be properly divided into the metalliferous and
non-metalliferous; none, he says, are necessarily destitute of ores, but, in some
mountain-chains, the fissures are too few in
number, or too inconsiderable in extent, to

allow the rocks to be metalliferous, be their substance what it may.

11. ON THE FOSSIL CONTENTS OF ROCKS.

It is said in the Wernerian theory[a], that there is a relation between the nature of fossils, and the age of the rock which contains them; that zoophytes, the lowest and most imperfect class of animals, were first formed; then, shell fish, and marine plants, of a nature unknown to the present world; afterwards, the same genera of animals as those we are acquainted with; and, still later, the same species. Land-plants are said to be more recent than any of these, and land-animals more recent still.

Whether Werner derived this idea from the study of nature, or from the perusal of

[a] See Von Buch's Norway, translated by Black, p. 47, and the note by Professor Jameson. Also in Transactions of the Royal Academy of Berlin, a Memoir by Von Buch, " Ueber das Fortschreiten der Bildungen in der Natur."

Telliamed, who imagined, that birds and
beasts sprung originally from the sea, and that
men and women themselves were only an
improved breed of fishes, I know not ; but
there seems as little reason to believe, that
a progressive amelioration took place in the
nature of the beings called into existence
previously to the deluge, as that a corresponding degeneracy can be attributed,
with any degree of justice, to those of
succeeding ages.

It is not correct to say, that zoophytes were the first born of animals : for
the genealogy of the nautilus is quite as
long as that of the madreporean polypus.

It is not always, nor, perhaps. generally,
true, that the fossils of the older rocks are
more unlike the productions of the present
world, than the fossils of the newer. The
fishes, whose forms are impressed on the
bituminous marl-slate of Germany, are not
more strange to us than those found in the
lias, the Portland stone, the Purbeck stone,
the chalk, or the rock of Monte Bolca,
and Oeningen ; nor the cacti, in coal-mines,
than the fruits of Sheppy Island, imbedded

in London clay ; nor the ancient corals of Derbyshire than the more modern ones of Berks and Wiltshire: nor the nautili of transition countries, than the ammonites, belemnites, hippurites, scaphites, baculites, and nummulites, which occur in rocks of a much later æra. Among the larger animals, one of the oldest, we are acquainted with, is the crocodile : now the fossil crocodile is certainly not more unlike an animal of the present world, than the megatherium, the palaiotherium, the anoplotherium, which are all found in rocks comparatively modern, or even than the mastodon, interred in alluvial soil.

It is said to be certain, that oviparous quadrupeds[a] had the start of viviparous. The latter have been found, hitherto, in regular strata, at Montmartre, and one or two other places ; but we cannot suppose, that all the viviparous animals, which existed before the deluge, were interred there only ; many must have existed in

[a] Cuvier, Discours Preliminaire, p. 68.

other districts, though their bones have not yet been discovered, or recognized, or perhaps have perished ; and who can venture to assure us, that the strata, in which they were enveloped, are not of more ancient date, than those which belong to the basin of Paris ?

The occurrence[a] of wood in flötz-trap has been adduced in proof of the newness of that formation ; but wood is found in every rock from the lias upwards : it is found also in the coal-measures; the immense tree preserved in the Museum at Dresden, is said to have been imbedded in porphyry.

The occurrence of deer's horns [b] would furnish a more convincing argument, if these horns are really found in trap.

An opinion has for some time past been entertained in this country, that every rock has its own fossils.

" Quarries of different stone," says Lis-

[a] Jameson's Mineralogy, vol. iii. p. 85.　　[b] Ibid.

ter[a], " yield quite different sorts, or spe-
" cies, of shells ; the cockle-stones of
" the iron-stone quarries of Addertone, in
" Yorkshire, differ from those found in
" the lead mines of the neighbouring
" mountains, and both these from the
" cockle quarry of Wansford bridge, in
" Northamptonshire ; and all three, from
" those to be found in the quarries about
" Gunthorpe and Beauvoir Castle."

The same writer followed the course of
the chalk marl over an extensive tract of
country by mere attention to its fossils.

Mr. Strange[b] traced the gryphus from
the lower part of Monmouthshire and
Purton Passage, through Gloucestershire,
Worcestershire, Warwickshire, and Leicestershire, occupying in these counties, as in
Northamptonshire, the lower parts under
the hills.

" The mountains of Glaris," says another
author[c], " produce nummulites, ammo-

---

[a] Philosophical Transactions, No. lxxvi.
[b] Archæologia, vol. vi. p. 36. for 1782.
[c] Journal de Physique, Introduction, vol. ii. p. 606.

" nites, crook-beaked oysters, and sea-
" shells of distant or unknown parts,
" imbedded in rocks of coarse limestone;
" whereas the adjacent mountain of Blut-
" tenberg, being composed of black slate,
" affords only the skeletons of fish."

After pointing out various circumstances
by which different strata may be discri-
minated, Mr. Calcot [a] says, " they are dis-
" tinguishable, still more remarkably, by
" the fossil bodies they contain, one layer
" abounding with one species of shells,
" another with a different; another layer
" containing bones and teeth of fishes,
" another corals of various kinds," &c.

Deluc [b] observes that " the fossils of chalk
" are different from those of other beds, and
" that gypsum also has its peculiar fossils."

Mr. Martin [c] assures us that " in Derby-
" shire the shells, corals, &c. which abound
" in the first limestone strata, are by no
" means frequent in the second, nor are the

[a] Calcot on the Deluge, p. 161.
[b] Journal de Physique for 1791, vol. xxxviii. p. 175.
and p. 181.
[c] Tilloch's Magazine for 1811.

" petrifactions of the second limestone com-
" mon to the first or third, though all these
" strata are evidently constituent parts of the
" same formation. In the coal soils of that
" county, the constituent strata of gritstone,
" ironstone, shale, &c. may all be charac-
" terized," he thinks, " by their respective
" vegetal fossils."

Mr. Smith is well known to have
embraced this idea at an early period, and,
in a table attached to his Geological Map of
England, has specified a variety of fossils,
by which the strata of England may, in
his opinion, be identified.

That the fossils contained in secondary
strata, are, of all empyrical and accidental
characters, the most useful, in enabling us
to follow the direction of these strata, no
one can dispute ; but their utility has been
greatly over-rated. Those who maintain
that formations are universal, and produce
every where the same fossils, must maintain, that these fossils are also universal ; in
other words, that every part of the earth
has been peopled, at the same period, by

the same animals, which, from the nature
of many of those animals, is absurd : for if,
in the pre-existing world, as in the present,
different animals inhabited different countries, then are fossils not universal ; but
each class confined to an area of greater or
less extent, and consequently, within and
without that area, the fossils of the same
formation will not be the same.

In South America belemnites and ammonites are entirely unknown.

The belemnite is said to be characteristic
of the chalk of France ; in the chalk of
Ireland it is common ; in that of England
rare. Many differences are observable
between the fossils of the Kentish Downs
and those of Wiltshire. The fossils of
Cherryhinton, in Cambridgeshire, differ
from both. The plants impressed on the
coal-slates of Somersetshire are very unlike
those found under similar circumstances
in Yorkshire. The rock at Maestricht is
said to belong to the chalk formation, yet
its fossils are peculiar, or nearly so, to that
spot. Of the five hundred varieties of fruit
found at the Isle of Sheppey, scarcely one

has been discovered in any other portion of the London-clay. The Tisbury
bed contains Coral at Tisbury only. Of
the fossils which have been described as
belonging to the lower marine formation
in France, very few have been recognized
in England.

If fossils were distinctive of the beds in
which they occur, they should severally occur
in one bed only ; now what fossil is there,
the range of which is so circumscribed?
the Palaiotherium and a few others : but
the bones of the Crocodile appear in
lias, in Stonesfield-slate, in cornbrash, in
Oxford-clay, in Kimmeridge-clay, in Port-
land-stone, in Tetsworth-clay, in green
sand, in the Maestrich rock, in London-
clay. The Crab, or part of that animal has
been observed in mountain-limestone, in
Stonesfield-slate, in iron-sand, in green-sand,
in chalk, in Folkstone-clay, in London-clay:
teeth and palates of Fish, in mountain-
limestone, in lias, Bradford-clay, cornbrash, Purbeck-stone, iron-sand, green-
sand, chalk, the sand of Reading, the
Sheppey-clay. Entomolites, in clay-slate,

transition-limestone, mountain-lime stone,
green-sand, chalk. Madrepores in transition-limestone, mountain-limestone, lias, in
the oolite of Bath, in that of Oxford,
in Stonesfield-slate, green and iron-sand,
in the unknown rock of Antigua, in agate.
Pentacrini, in the limestone of Ben-
thal-Edge, in lias, inferior oolite, Bath-
oolite, Stonesfield-slate, cornbrash, Kimmeridge-clay, chalk. Patellæ in the Walton
beds, Bath-stone, transition-limestone.ᵃ
Echini in mountain-limestone, red-marl,
lias, Bath-stone, Stonesfield-slate, iron-sand,
green sand, chalk, flint, &c.

Why should I extend this catalogue by
introducing oysters, ammonites[b], anomiæ,
common to almost all formations? Enough
has been said to shew, that, as in different
parts of the same bed we find different fossils, so also in different beds we find the
same fossils.

[a] Von Buch's Norway, p. 420. of the Translation.
[b] At Chamont, departement de l'Oise, ammonites are
found even above the Calcaire grossier.

Mr. Farey admits [a], that different rocks produce fossils of the same genera, but thinks the species characteristic; if that opinion be correct, it follows, that, as often as a new rock was deposited, all the then-existing species of shell-fish were destroyed and new species created, a circumstance by no means probable; let us look to the fact.

The Grignon-bed affords at least eight species of patella and tellina, ten of venericardia and turritella, twelve of melania and ampullaria, fifteen of mitra, bulimus, and cytharea, eighteen of ostrea and murex, twenty-five of pleurotoma, thirty-three of fusus, sixty of cerithium : it may afford still more, but these have been described.

Is the same species of fossil confined to only one rock? sometimes it is; nay more; it is confined to a small portion of that rock, perhaps to a single quarry; but more frequently the same species occurs in rocks produced at very different æras:

[a] Tilloch's Magazine, vol. xxxix. p. 89.

thus, the ammonites jugosus is found in oolite
and coal shale; the modiola depressa in London-clay and lias; the perna aviculoides in
the clay of Shot over and the aluminous rock
of Whitby; the terebratula lateralis in the
fuller's-earth of the Cotswold, and mountain lime-stone.

If the genera and species of fossils do not
furnish us with the means of determining
the relative age of rocks, it may be supposed, that the existence, or non-existence
of fossils, prove these rocks to have been
severally formed at an earlier or later
period than that in which organic beings
were formed. The grand division of rocks
into primary and secondary[a], is founded
chiefly upon this distinction.

Even here however, difficulties present
themselves.

The siliceous-limestone and millstone-
rock of Paris which contain shells, and
those which do not contain them, closely

resemble each other, not only in mineralogical character, but in position.

The shells on the summit of Snowdon
prove that animated beings existed at
the birth of the rocks composing that
mountain, and we are far from having
ascertained that any other rock can boast
a higher antiquity. The greenstone of
Tobago, which has all the mineralogical characters assigned to primitive greenstone,
also contains shells. In the museum of the
East India Company is preserved a similar
specimen with shells brought from Hindostan.

On the other hand, how many are there
of the more modern strata in which no
fossils have been discovered! If the absence of fossils does not prove, that gypsum,
rock-salt, grey-wether-sand-stone, &c. were
formed before the birth of organic matter,
how does their absence prove, that granite,
sienite, porphyry, &c. were formed before
that epoch ?

A very slight acquaintance with nature
is sufficient to convince us, that the presence or absence of fossils, in a rock,

depends less on the period of its depo-
tion than on the nature of its materials.
In the older formations, calcareous beds
often contain fossils, while the argillaceous are destitute of them. In Yorkshire
and Cumberland, the limestone abound
in fossil shells ; but the hazels, plates, and
toadstones, seldom, if ever, contain them.

Some writers have supposed, that crystallization is unfavorable to the existence, or
rather the preservation of organic matter,
and others have doubted, whether the presence of organic matter be not fatal
to crystallization ; but the red-marl is
not crystallized, though barren of shells ;
nor do we discover a more crystalline
structure in the varieties of argillaceous-
slate in which fossils are wanting, than in
those varieties in which fossils abound.

Generally speaking, vegetable remains
are most common in argillaceous beds,
and animal in calcareous.

The presence of fossils in some strata is
certain; their non-existence in others must
always be problematical. How rarely,
where bones are found in the best preservation, do we find enough of them to form

a complete skeleton ! Is it to be supposed
that the animals of which these constituted a part, were *monstra per defectum ?*
Is it not probable, that chalk which contains
so many teeth and palates of fish, contains
also other parts of the animals to which
these teeth and palates belonged?—and, if
so, that organic matter exists in many
siuations, in which its existence has not
been observed or suspected ?

ESSAY VII.

ON THE HISTORY OF STRATA, AS DEDUCED FROM THEIR FOSSIL CONTENTS.

THAT fossils are what they appear, and not nature's counterfeits, is a doctrine on which there is no longer any difference of opinion : it is remarkable, however, that, amid the vast number of fossils which have been discovered, there is scarcely one exactly similar to any existing plant or animal. All we can say then of fossil productions is, that they are analogous to recent ; but, of analogy there are infinite degrees, some so loose that to discover them is an indication of genius, others so striking that it would be a proof of dullness to overlook them. It is essential, therefore, in reasoning from recent to fossil bodies, that we should be aware of the *degree* of analogy, that we should know what are the fossil individuals that correspond with recent individuals, of fossil species with recent species, of fossil genera with recent genera; for on the *degree* of analogy depends the degree of evidence.

By neglecting this consideration, by identifying objects which are only analogous, by seizing on resemblances and overlooking distinctions, Naturalists, who, from the extent of information they possessed in their respective departments, might have rendered to geology essential service, have too often endeavoured to foist upon it conclusions the most wild and extravagant that can be imagined. Of an hundred and five different species of fish[a] enumerated as belonging to Monte Bolca, thirty-nine are said to have come from the Asiatic seas, three from the African, eighteen from those of South, eleven from those of North-America. What follows? that the fishes abandoned their native seas to meet at this spot, or that they were transported thither simultaneously over land or through the air? This is not likely. Shall we say then, that the fishes remained true to

[a] Saussure, § 1535.

their respective seas, but that the seas migrated ? This is not likely. Two other suppositions remain ; one, that Conchology is fallible in its decisions; the other, that the propounder of the doctrine was ill-grounded in Conchology : whether either or both of these suppositions be true, I leave to the judgment of those, whose judgment upon such a point is more valuable than mine.

It has been stated[a], that in every coalmine the fern of America is blended with the palm of Africa, and the Asiatic bamboo. What follows ? Did Asia, Africa, and America travel to England, or England to Asia, Africa, and America ? How happens it that one and the same deposition went on quietly and uninterruptedly in spite of the locomotion ? How happens it that our floating and fortunate island, enjoying at the same moment the advantage of opposite climates, produced plants both of tropical and polar regions? If these things are incredible, there is reason to apprehend that we may be misled, as well as instructed, by Botanists.

[a] Dolomieu, Journal de Physique, vol. xxxix. p. 399. See also Playfair's Illustrations, p. 178.

In their most perfect state, all the sciences would be strictly in unison. Nothing can be true in natural history which is impossible in natural philosophy. It is not surprising, therefore, that modern observers should find the observations of their predecessors inaccurate, when opposed to the conclusions of every other science. Brocchi and Reniere think that many of the shells, found at the foot of the Apennines, which have been deemed peculiar to the East Indies, Africa, America, and different seas in Europe, now exist in the Adriatic. Poli has found, in the bay of Naples, many shells, and Maratti, many zoophytes and madrepores, which are usually considered to be the productions only of distant seas. These discoveries, however, do not take away from the marvellous ; they only keep it in the back-ground. The resemblance between the Sub-Appennine fossils and the recent shells of the Mediterranean and Adriatic, if established, is a coincidence, extraordinary indeed, but fortuitous ; for it is evident, that these fossils were interred at the period in which strata were deposited,

at a period when the relative positions of
land and sea were different from what
they are at present; when perhaps [a] the
Adriatic and Mediterranean were not in
existence.

———

The alternation and occasional intermixture of sea-shells with those of fresh-
water, is common to all the secondary
strata, and not unknown in the transition.
The greywacke-slate of the Harz contains encrini and reeds. Sea-shells, accompanied by impressions of fern, are observable
in the dunstone of Ludlow and South
Wales. Coal-shales and nodules of iron-
stone exhibiting casts of fresh-water muscles,
are often interposed between the coralline
limestones of the northern counties. The
monitor, which occurs in the copper-slate
of Thuringia, is associated with fresh-water
fishes and marine shells. The lias affords
ferns, nautili, and crocodiles; the slate of
Stonesfield, remains of birds, beasts, and

* Vide supra, Essay, 2.

marine animals. The Petworth and Purbeck marbles, containing a species of
paludina, alternate with beds of sand-stone,
charged with marine univalves and bivalves;
fruit and leaves are found with marine
exuviæ in chalk. The clay at Sheppy
Island, abounding in sea-shells, is reported
to yield no less than five hundred varieties
of fossil fruit; fresh-water shells intermixed
with marine have been observed, also, at
Barton Cliff, at Brentford, and other spots
near London in the same bed. The alternation of fresh and salt-water productions
at Headen in the Isle of Wight, and in the
corresponding strata of the basin of Paris,
is notorious. At Guespelle, at Pierre-
Laie, at Grignon, &c. [a] sea-shells are intermixed with fluviatile. At Montmartre [b] the
gypsum exhibits animals of land, air, and
water; the middle beds [c] of that rock
contain fresh-water shells; the upper and
lower, marine.

[a] Journal de Physique, vol. lxxvii. p. 362.
[b] Ibid. vol. lxxi. p. 391.
[c] Ibid. vol. lxxvii. p. 365.

In the area included by a line drawn from Mayence through Frankfort and through Hockheim back to Mayence similar alternations are observed of fresh and salt-water productions. At Monte Bolca, Pappenheim, and Oeningen, impressions of fishes occur with land plants, and at Morte Pulgnasco[*], the bones of the elephant and rhinoceros are mingled with those of cetaceous animals.

How these extraordinary alternations and intermixtures are to be accounted for, and whether they are attributable, in all cases, to one and the same cause, it is difficult even to conjecture. In the basins of Paris and the Isle of Wight, the only districts in which the subject has been properly investigated, it has been thought the most easy method of solving the problem to imagine alternate inroads and retreats of the sea, coupled with the occasional existence of fresh-water lakes.

This hypothesis, however, is open, I

[*] Journal de Physique, vol. xxxix. p. 399. Vol. lxvi. p. 105. Vol. lxvii. p. 81. Vol. lxxx. p. 50.

conceive, to insurmountable objections. The supposed fresh and salt-water beds, are identical in substance, and comformable in position; there is no mineralogical difference between the beds of gypsum, which contain cerithiæ, and those which contain cyclostomata, lymnæœ and planorbes ; between the marine-lime-stone and the fresh-water limestone, the marine-grit and the fresh-water grit. Is it possible, that, the depositing menstruum having changed, the matter deposited should not have changed also ? or that, a sea having retired before a lake, or a lake having been overwhelmed by a sea, no trace of such catastrophe should be visible any-where on the then, and still-unconsolidated materials, which furnished the scene of action ?

Is the distinction between fresh-water and salt-water shells so strongly marked that they cannot be confounded ? The common test is the thickness of the shell, but sea-shells are by no means uniformly thick, as we see in the oyster, &c., nor those of lakes and rivers uniformly thin. In a series of bulla, patella, pecten, pinna,

argonauta, &c., it is easy to find shells
as delicate and fragile, as those which are
usually contained in rivers or lakes.

I am not aware of any other character,
by which a naturalist can distinguish *à
priori* a fresh-water shell from one inhabiting the sea.

————

Shells have been divided [a] into littoral
and pelagian, but it is difficult to say in what
depth of water any fossil was imbedded.
Pentacrini, ammonites, oysters, crocodiles,
together with a quantity of wood, are found
in the same rock. Was the sea, which
deposited that rock, deep and shallow at
the same moment?

It seems admitted by all naturalists who
have considered the subject that fossil
productions, whether vegetable or animal,
found in northern latitudes, have a general
resemblance, a family likeness to vegetables

• Rouelle and Lavoisiere, Journal de Physique,
vol. lxxv. p. 446.

and animals, found recent in southern
latitudes; but that the converse of this
is not true.

The elephant of the Lena river presented
an entire skeleton covered with skin and
hair; the character of the hair by no means
justifies the common opinion that this
animal was the native of a southern climate.

Its remarkable preservation is owing probably to the ice by which it was surrounded; and if so, the ice must have
formed nearly at the same time at which
the elephant died; thus we obtain a clue
to the temperature which then prevailed
in that part of the world, and a proof
that the same degree of cold has prevailed
there ever since. These facts are of great
importance to the speculative geologist,
but, as yet, thoroughly inexplicable, and
it is better to avow our ignorance than to
display it.

* x

ESSAY VIII.

ON *MINERAL VEINS.*

DR. KIDD observed one morning a narrow crevice running along his path. A few days afterwards he found this crevice widened, and, opening into it on either side, many lateral crevices, which at the further extremity became gradually narrower till they were lost.

It is probable that fissures in general have been produced, as these undoubtedly were, by shrinking, and are coeval with the consolidation of strata.

A crystal of calcareous spar will break most readily in laminæ. A broken crystal of² quartz will be generally speaking conchoidal, and of corundum ragged; so each uncrystallized body has a more or less determinate fracture. Blocks of granite

7

and limestone assume one figure; of basalt, another; of slate, a third: they all break in that direction in which the cohesion between their several particles is least: the form of the block depends therefore on the nature of the substance, and the form of the fissure depends on the form of the block.

According to the laws by which the cohesion of its particles is regulated, a stratum in shrinking will divide into large blocks or small. The pedestal of the statue erected to Peter the First was hewn out of a block which weighed 1500 tons. How different its dimensions to those of the fragments which we see re-united in a specimen of ruin-marble from Florence, or jasper from Bohemia!

Fissures may also be produced or enlarged by the contraction of an adjoining mass, by the shock of an earthquake, or by failure of support, owing to the erosion of subterranean waters.

The fissures in Septaria can have been occasioned only by contraction, but contraction hardly occasioned the fissure in Yorkshire, which extends from the eastern moorlands to the western, and disturbs indiscriminately in its progress all the beds which are interposed between the oolite series and that of the mountain limestone.

In the north of England, both varieties of fissure expand in calcareous beds, contract in siliceous, and almost close in argillaceous; but the rule does not hold good universally. The rocks at Adersbach, in Bohemia furnish a convincing proof, that fissures of sandstone may be as considerable as those of limestone.

The widest vein that occurs to me is in the Isle of Arran; it measures 160 feet across. There is one at Lauterbrun, in Swisserland, which measures 140 feet.

The inclination of fissures is regulated in some degree by the nature of the intersected stratum. In the mining districts of Yorkshire, Cumberland, and Durham, they are said to be nearly vertical in calcareous beds, in argillaceous nearly horizontal, in siliceous oblique.

At Cooper's dyke head, near Alstone, the sun vein cuts the coal-sills, white hazel, and little limestone vertically; it dips four feet in six where it traverses the plate; runs horizontally for five feet along the surface of the great limestone, and then cuts it vertically.

Extensive fissures are for the most part accompanied by a subsidence of the adjacent strata, and, generally speaking, the amount of the subsidence is more or less considerable in proportion to the size of the fissure. Subsided beds sometimes preserve their parallelism; instances of this may be seen at Fairhead, and on the west of Aust passage, or, on a smaller scale, in specimens of Florence or Cottam marble. More commonly their parallelism is lost.

Where a subsidence has taken place, the strata on the hanging side are said to be lower than those on the lying side; in other words, the subsided strata form the roof of the vein. Exceptions to this rule occur at Westbury Cliff on the Severn, and at Balahulish in Scotland.

Fissures which cut the strata obliquely have been observed to produce more dislocation than those which cut them vertically.

The quantity of dislocation occasioned by a fissure, will differ at different places.

Fracture and subsidence do not necessarily produce fissures. In the collection of Signor Castellini at Schio, a specimen of calcareous slate from Monte Bolca on one side exhibits the impression of a fish's head, on the other side the impression of its tail. In this case, the upper strata have merely slid over the edges of the lower. Of the same nature is the extraordinary slip at Handfast Point in Dorsetshire.

ON MINERAL VEINS.

Veins are chasms filled with mineral matter; if with stony matter, they are often called dykes; if with tin or copper, lodes.

If they intersect the strata they are called rakes; if they lie between them

311

pipes, flats, or dilated veins. The latter
are comparatively rare.

Where a section passes near the line of
junction of two rocks, the surfaces of
which are curved or irregular, portions of
the one will be seen protruding into the
other. Appearances of this kind have frequently been observed at the junction of
granit or porphyry with killas, and the protruding portions have been improperly
called veins. There are real veins however
of both these substances.

A vein that lies between the strata in
one place, will often intersect them in
another. This has been observed particularly of whin dykes [a], but it is no less
true in regard to metallic veins.

Pipes have excited most attention in the
north of England, where they prevail in
the Tyne-bottom-limestone the scar-lime-
stone, and the great limestone of Alstone

[a] To the examples adduced by Mr. Jameson and Dr.
M'Culloch may be here added some remarkable ones
at Pallis in the county of Limerick, at Carlingford in
the county of Louth, at Scormore in the island of
Rum, and along the shore of Ardnamurchan.

x 4

Moor, not in the plates or sandstones.
At Virgin-mine in Wenzley-Dale, are two
pipes, which have been followed to the
distance of nearly a mile. Three have been
worked at Corser and Tyne-head. At the
Old Carr mine at Nenthead, they have been
very productive.

At Polgooth[a] and Carnmeal mines in
Cornwall a floor of tin was worked twelve
feet in breadth, but without the determinate walls usually observable in regular
veins.

————

Veins differing in their contents often
intersect.

Veins agreeing in their contents take
different directions in different countries.

In the same country similar veins have
often, but not always the same direction.
At Schemnitz, the veins are said to run
north and south; at Scharfenberg, in
Saxony, east and west; at Freyberg, north

* Geol. Trans. vol. ii. p. 129.

and south, or north-east and south-west In Wicklow, the lead-veins run east and west ; in Bretagny, north and south.

It was observed by Owen, as early as the year 1595, that the veins in England run from east to west, and this is still the prevailing opinion.

In Cornwall it is thought, that the tin lodes point rather to the north of east, and the copper-lodes rather to the south ; but I believe the deviation rarely amounts to twenty degrees. At Wendron, and at Huel Jubilee near Padstow, are copper-lodes, which, if I am rightly informed, run north and south.

The direction of the principal lead-veins in Cumberland and the adjoining counties, seems also to be east and west, but the smaller veins run in every direction. The most considerable lead-vein known in Devonshire, the Bere-Alston load, runs nearly north and south, as do most of the lead veins worked in Cornwall.

The direction of veins [a] is no where determined by the direction of hills; but the nature of the rocks which compose these hills may have influenced the direction of both. Generally speaking, veins were formed before the hills and valleys which now vary the surface of the earth.

Veins are sometimes composed of one substance only, and that nearly the same chemically speaking, as the rock in which they occur; as calcareous veins in limestone, whin-dykes in whinstone; but their contents are often extremely multifarious. Almost all the simple minerals known, are found in veins occasionally, and the greater number exclusively.

Veins produce different substances in different parts of their course.

The Stephani-lode, at Schemnitz, yields lead and gold at one place, silver at another. At Kremnitz the contents of the main lode

[a] See Essay II.

change from gold and hornstone to clay
and pyrites. The Old-Gang affords lead in
Swaledale, but at Muker and Middleton
Tyas, copper. The vein at Welhope, is
free from sulphate of barytes at one shaft;
at a little distance, sulphate of barytes is
so abundant that it measures four feet in
width. The Rampgill vein, celebrated for
fluor at Allenheads, is destitute of it at
Coalcleugh. The Huel-Virgin lode in Cornwall, has been followed to the distance of
two miles; but the arseniates of iron and
copper are found in a very small portion
of it. Plush-copper and uranium do not
accompany the Huel-Crebor lode all the way
from Tavistock to the Thamar river, but
only at Gunnislake.

Changes of this nature are most common
where veins pass from one rock into
another.

Bergman has noticed this at Kongsberg in
Norway, and Picot de la Peyrouse[a] attributes the irregularities and interruptions
of the metallic veins in the Pyrenees to the

* Journal des Mines, vol. vii. p. 5

frequent alternation of the rocks through which they pass. [a]

Calcareous veins are most frequent in the secondary rocks ; quarz veins in the primitive.

On the east-side of St. Michael's mount, in Cornwall a vein cuts both the granit and killas ; in the granit the substance of the vein is shorl ; in the killas, quarz.

Another vein on the same island produces, or does not produce, tin-ore, as it passes through granit or through killas.

At Homebush, near Callington the lode is of tin, where it traverses granit ; of copper, where it traverses killas.

At Tresavean, in Cornwall the lode is very poor in killas, but rich in granit. The reverse takes place at Huel Unity.

In the lodes worked in granit, there is in general a greater variety of substances than in those worked in killas ; this may easily be verified at Huel Gorland, Huel Damsel, Carrarack, and Gunnislake which are

---

[a] Trebra has remarked, that less argillaceous strata are fertile of precious metals than granit and porphyry. J. des M. vol. iv. p. 72.

worked in both. The lode at Huel Unity
bears fluor, only where it traverses granit.

I have been told by Cornish miners, that
the tin is generally better in elvan or porphyry than it is either in granit or in
killas.

In the north of England the most
valuable mines are all situate in limestone;
the hazels are metalliferous in a much
smaller degree, and in the plate or beds of
indurated clay the veins carry for the most
part clay only. If lead has ever been
found in plate, it is only where, as at Thorn-
gill and Blagill, the plate varies from its
usual character, and passes into chert.

When, in consequence of a fault, limestone
and plate face each other, and there is a
rider of limestone, the lead generally lies
between the rider and the lime-stone, very
rarely between the rider and the plate.
The richest veins are said to be those of
which both cheeks are alike.

The Cornish mines appear to be most

productive at the junction of granit and killas : those of Derbyshire and Flintshire, at the junction of limestone with limestone-shale. Raumer has observed an enrichment of the Silesian veins at the junction of different rocks. Ferber remarks that the mines at Agarth near Belluno, at Feltrino, at Schio in the Vicenza territory are all, like those in the Bannat, situate at the junction of slate and limestone.

From these considerations we might naturally infer that the nature of veins is influenced by that of the rocks they intersect ; this however is doubtful, for

1. Rocks, which abound in any given veins in one country, contain no such veins in another. Greenstone-slate, for instance, is metalliferous in Silesia and Sweden, not in Great Britain. About Mohorn in Saxony clay-slate is not metalliferous; yet the mines at Schneeberg are all in clay-slate.

Of the limestone tract in North Wales a very scanty portion only produces lead-ore. The great limestone of

Alston is more metalliferous towards the
west than towards the east, and more so
near the surface than far beneath it. The
upper limestone in Derbyshire is the most
metalliferous, though it agrees with those
below it, both in chemical and mineralogical
character.

2. Similar substances occupy the veins of
dissimilar rocks.

A vein consisting of galena, black-blend,
arsenical-pyrites, iron-pyrites, copper-py-
rites, and a small quantity of brown-spar,
occurs at Freyberg in gneiss, near Mohorn
in porphyry, near Muntzig in clay-slate. A
vein consisting of galena poor in silver,
barytes, and fluor, at Freyberg traverses
gneiss, in Derbyshire limestone.

3. Dissimilar substances occupy the veins
of the same rock. In Cornwall the veins
worked in granit are chiefly of tin and
copper ore ; in Bretagny and Wicklow, of
galena.

The sienit of the Isle of Cyprus affords

copper; that of Hungary, gold and silver; that of Thuringia, iron.

The metallic ores are almost always associated with vein-stones, and ores of different metals belong to the same vein. Without any change of the stratum, we often find a change in the nature of the vein, either in its longitudinal bearing, as at the Old-Gang which produces lead in Swaledale, but copper at Muker and Middleton Tyas, or in the direction of its descent, as the lodes of Cornwall, in which tin lies over copper or alternates with it. *

4. Substances occurring in veins, are for the most part extremely different, in a chemical point of view, from the rocks traversed by them. Thus, stony bodies are intersected by metallic ores. Fluor and barytes are ordinary constituents of veins, as constituents of rocks unknown. No one can imagine a chemical relationship between granit or killas, and the endless variety of simple minerals found in the veins of Cornwall.

* This happens at Tincroft.

And here I must be permitted to observe, that if it is no longer allowable to believe in the philosopher's stone, neither are we at liberty to believe, as many writers are inclined to do, that mineral veins owe their origin to electricity, galvanism, or some unknown cause similar to these. No analogy will support so wild an hypothesis, as that one, or all of these agents should have converted particular portions of rock not merely into a different rock, but into a variety of spars and ores, differing alike from it, and from one another.

We are warranted then, I think, in concluding, that, if the nature of the rock has any influence on the nature of the veins that traverse it, its direct influence is but extremely feeble. Its indirect influence is more considerable, because on the nature of the rock depends the extent, and, in some degree, the direction of the fissures.

It has been supposed that the toad-stone beds in Derbyshire cut out the metallic veins, and are consequently more modern.

The fact is not true, at least invariably. At the High Rake near Tideswell a con-

Y

siderable vein of lead and spar has been followed through the toad-stone for forty-five
fathoms in depth. Lead has been worked
in the same situation near Castleton in
Derbyshire, at Garrigill Gate in Cumberland, and at Kady in the county of
Armagh.[a]

The way-boards which divide the mountain lime-stone into beds, the shale which
covers it, the chert imbedded in it, the
plate alternating with it, often present to
the metallic veins a barrier as formidable
as the toad-stone does. In the north of
England it is considered an extraordinary
circumstance that the lead vein at Thorn-
gill and Blagill penetrated all the " sills
and plates" up to the very turf.

On the coast of Fifeshire I have seen a
vein traverse the subjacent and superincumbent sand-stone, but not the lime-stone
which intervened.

Near High Pike[b] in Cumberland a vein,
which continues its course uninterruptedly

[a] Whitehurst Inquiry, p. 264.
[b] Phil. Mag. for Jan. 1816.

through sienit to a distance of three miles, ceases suddenly where it leaves the sienit, and reaches the slate.

At Gersdorf [a] in Saxony, the veins worked in gneiss do not traverse the superincumbent clay-slate; those worked in gneiss at Nicholasberg in Bohemia do not traverse the superincumbent porphyry.

Veins of asbestus, which abound in the drawing-slate of Switzerland, do not penetrate the lime-stone with which this slate alternates.

Whether in all the instances here adduced the vein has really been cut off I am incompetent to decide. My belief is, that in some cases the interruption is real and complete, and that it is so not only where veins quit one rock to traverse another, but in their passage through the same rock. Instances of this may be seen at Glenrosa in the Isle of Arran, and at Bunessan, in the Isle of Mull. On a smaller scale interrupted veins, both of quartz and lime-stone, are well exposed

[a] Reichetzer's Geognosie, p. 232.

Y 2

along the slaty cliffs that project into the sea at Ilfracombe. Need I bring to the recollection of my readers the interrupted vein described by Dr. Mac Culloch, in a mass of lime-stone at Waltham Abbey?

In Derbyshire, it would appear from the observations of Mr. Farey [a], that the smaller veins are effectually stopt by the toad-stone, but from the larger veins a crack or rent generally extends some distance into the toad-stone, both above and below, growing narrower, and often branching into different small cracks or strings as it proceeds.

In interrogating miners on subjects of this nature, it is useful to bear in mind that in their language a vein is said to be cut off, when from a diminution of quantity or deterioration of quality in the ore, it ceases to pay the expense of working. At Schemnitz the Stoplitzhofer-grund is said to cut off all the veins; it only impoverishes them.

If it were true that veins were destroyed, or shifted as often as they came in contact

---

[a] Survey of Derbyshire, p. 245.

with toad-stone, and that in the latter case
the part shifted were of the same dimensions and of the same nature as that not
shifted, we might, with some plausibility,
refer the shift or destruction to an inroad of the toad-stone between the strata
after the veins were formed, but then we
must in fairness refer to a similar cause
all similar phænomena. It is evident,
however, that no inroad of toad-stone can
have occasioned a ramification of veins,
a reduction of their size, or an alteration
in their contents. These are circumstances which admit no rational explanation, either by the volcanic or the plutonic theory: they appear to depend on the
extent and direction of the intersecting
fissures.

The same reasoning may be applied to
veins said to have been shifted by other
veins. Subsidences have taken place of
veins, as well as of strata, but I am persuaded that many veins are supposed to
have been broken, which were never continuous, and many shifted which retain at
this day their original position.

To establish the original continuity of veins on the opposite side of a cross-course, it is not sufficient to shew an agreement in one spot; we must shew it in many. If by an imaginary shift of a stratum we can bring together, as at Huel Peevor, in Cornwall, the mouths of several veins, then, indeed, we are justified in supposing that such shift has taken place since the veins were filled; but we are not to conclude, because the ends of a road or river on different sheets of a map would coalesce, that the sheets were intended to be so joined; the valleys of the Lea and the Cray both open into that of the Thames, the one from the north, the other from the south, but it does not follow that they were continuous, till the Thames destroyed their continuity.

Some veins are said to have been impoverished on one side of a cross-course, and enriched on the other. May not the parts impoverished and enriched have been parts, not of the same vein, but of different veins?

At Schemnitz is a vein which bears lead and quartz, till crossed by the Wolfgang lode, after which it is said to bear only clay;

but how is it ascertained that the vein, which bears clay only, is the same as the metalliferous?

In the neighbourhood of Alstone two parallel veins are said to be shifted by a cross-course in opposite directions.

Near Redruth two copper lodes, those of Huel Virgin and Huel Maid, meeting obliquely, are shifted by a cross-course, the one to the left, the other to the right.

Where a lode is really shifted by a cross-course, the latter will often contain portions of the former, and the direction of these indicates the direction in which the lode is shifted.

To identify two lodes in the same mine is often a work of considerable nicety, and yet I have met with practical men, who think that the veins which bear calamine in Mendip are the same as those which bear calcareous iron ore in the forest of Dean; who identify the whin-dykes of the Hebrides with those of Antrim; and recognize in the Isle of Man the same veins which have proved productive in Wicklow and in Swaledale.

That in this country the east and west veins are for the most part shifted by the north and south, is a common opinion among miners both in Cornwall and at Alston Moor. In both districts, however, we meet with numerous exceptions. At Hawkside in Harwood a north and south vein is shifted by one running east and west. The great cross-course in Cornwall, which stretches from the English to the Bristol channel, shifts every thing it meets with, except a small fluchan, by which it is itself shifted. The great copper lode of Cronedale in Devonshire, repeatedly shifted by north and south cross-courses, shifts one of these at Huel Luscombe. A similar circumstance attends the lode at Bere Alston in the same county.

According to Werner, the different substances of which veins are composed appear in determinate order on either side.

Of such arrangement I have seen repeated examples, particularly in whin-dykes,

but the exceptions I believe are so numerous, as to do away the generality of the
rule. In the veins which I have had an
opportunity of examining in Derbyshire
and Cumberland, barytes and calcareous
spar change places continually. It is not
easy to discover a lode in Cornwall, where
tin and copper lie in the orderly manner
which Werner described. At Glencloy [a]
in Arran, a vein, of which the centre is
whin, has one of its sides composed of
brecchia, the other of siliceous sand-stone.
At Tormore in the same island, the one
side of a dyke is basalt, the other porphyry.

———

" By all the information I could ever
" procure," says Hutchinson, " I cannot
" perceive there is any instance of a dis-
" position of ore in Hungary, Saxony,
" Mexico, Achin, or elsewhere, of which
" we have not some example in England,
" so that he who is thoroughly informed of

[a] Jameson's Scottish Islands, p. 27.

" the condition of things under-ground in
" this island, is qualified to form a judge-
" ment of them all round the globe." [a]
We here recognise the doctrine afterwards promulgated at Freyberg, of vein-
formations. According to that doctrine,
the materials of veins are diffused as universally as those of strata, and if a lead
vein in Derbyshire afford the same vein-
stones as a lead-vein in Saxony, a common
origin is ascribed to both.

" If," says Humboldt[b], " we had an
" accurate account of the four or five thou-
" sand veins in New Spain, that are either
" now working or have been worked within
" the last two centuries, we should, no
" doubt, perceive in the contents and struc-
" ture of these veins such analogies as
" would prove their simultaneous origin.
" We should find their contents in part
" identical with those of the Saxon and
" Hungarian veins."
I am disposed to pay due deference to

[a] Hutchinson's Works, vol xii. p. 368.
[b] Journal de Physique, vol. lxi. p. 273.

these authorities, but if, as has been shewn, the same vein is not uniform in its products or dimensions to the extent of a few perches or fathoms, surely the resemblance of these products at one spot to the products of another vein at another spot is not sufficient to prove that they agree in their character throughout; nor if their identity were established, should we have reason to ascribe to both a common age or a common origin, unless it be absurd to suppose that similar veins, like similar strata, may have been formed at different æras.

It is remarked by Patrin [a], that a zone of copper, lead, and silver ore stretches nearly in the same latitude from England to the eastern extremity of Asia, and thence to North America. In this zone are comprehended the mines of Ireland, England, France, Germany, Hungary, and Transil-vania, the Altai, the banks of the Amour, the shores of Kamscatka, and the blue mountains of America. To establish this

[a] Journal de Physique, tom. xxxviii. p. 299. see also Genetti.

doctrine it would be necessary to prove the non-existence of copper, lead, and silver in other parts of the old world. Were Africa and Asia sufficiently explored, similar zones might probably be drawn in other directions with equal propriety.

The junction of veins with the beds they traverse is often attended with curious circumstances, which it is difficult to account for on chemical principles, though in explaining them it is no less difficult to resort to any other. What is *moor-stone* at a short distance from the lodes, will often become *grauan* in their vicinity, or the grain of the granit will be altered, or it will become [a] slaty ; or the feld-spar will change into adularia [b], or one of its three ingredients [c] will disappear, or a fourth be added.

[a] I have seen this on the side of a vein of micaceous iron ore near Irton, in Cumberland.

[b] Goatfield, in Arran.

[c] Aberdeenshire, Wicklow, &c.

Such anomalies are particularly numerous and striking by the side of whin-dykes.
Along the line of contact common sandstone is sometimes converted into jasper[a]
or lydian-stone; chalk into granular marble[b]; coal into coak or plumbago[c]; clay-
slate into hornblend-slate. [d]

The most common effect of whin-dykes,
appears to be that of hardening the strata
adjoining them, but the effect is by no
means constant; at Glencloy in Arran,
and at Mr. Kennedy's quarry near Belfast,
the sides of the whin-dyke are of clay.

The relative age of veins is a subject on
which Werner bestowed considerable attention, but respecting which our knowledge
still remains extremely confined. It is reasonable to suppose that the oldest veins are

[a] At the blaue Kuppe at Eschweiler, in the department of Forets; on the north-west side of Stirling Hill,
at Portree, in the Isle of Sky, and Salisbury Crags.
[b] Antrim and Rathlin Island.   [c] Ayrshire.
[d] Glen Tilt. See Geol. Trans. vol. iii. p. 270.

of a later date than the consolidation of the
beds in which they occur, because the fissures which they occupy appear to have
been occasioned for the most part by the
shrinking of these beds during consolidation. We may fairly conclude also, where
different beds are traversed by a vein, that
the vein is of posterior date to the most
modern of these.

Of two veins that cross each other, the
intercepted must be the oldest ; we should
take care, however, to apply this rule
only where the vein is really intercepted.
Mr. Playfair tells us that in Cornwall granit
veins intersect the metallic, and are remarkable for producing shifts in them: no
such instance has come within my observation.

According to Werner[a] veins which have
no way-board are nearly contemporaneous
with the rock in which they occur, but the
value of this remark is done away by
another remark of the same author, viz.
" that a vein is seldom united to a rock

[a] Werner on Veins, p. 137 of the translation.

" through its whole course ; this takes place
" only in certain parts." Stone-dykes, I believe, have rarely a way-board ; the whin-
dyke at Cockfield, which has none, traverses
mountain lime-stone and lias ; is it contemporaneous with both ? [a]

The way-board depends, if I mistake
not, much more on the nature of the vein
than on the period of its birth.

It is maintained in the Huttonian [b]
theory, that metallic veins were not formed
till after the secondary strata ; the foundation of this doctrine I am at a loss to discover. Tin, molybdœna, tungsten, wolfram,
uranium, bismuth, and titanium seem to
belong exclusively to the rocks of primitive
countries. The successive coats of agate
and stalactite are in some measure analogous to veins, and these as well as the veins
which we find in septaria, and nodules of
argillaceous iron-stone, are evidently the
effect of secretion or infiltration.

The Huttonian hypothesis, that veins

[a] Werner on Veins, p. 91 of the translation.
[b] Illustrations, p. 123.

have been filled from beneath, appears to me perfectly gratuitous ; and the Wernerian hypothesis, that they have been filled from above, though it derives some support from the circumstance of trees and rounded pebbles having been occasionally found in them, is irreconcileable with the alternate opening and closing of veins which we have shewn to take place not unfrequently on their passing from one rock to another.

FINIS.

Printed by Strahan and Spottiswoode,
Printers-Street, London.

ADDRESS

DELIVERED AT

THE ANNIVERSARY MEETING

OF THE

GEOLOGICAL SOCIETY OF LONDON,

On the 21st of FEBRUARY, 1834;

By GEORGE BELLAS GREENOUGH, Esq., F.R.S.

PRESIDENT OF THE SOCIETY.

LONDON:

PRINTED BY RICHARD TAYLOR,

RED LION COURT, FLEET STREET.

1834.

ALERE FLAMMAM

ADDRESS, &c.

GENTLEMEN,

YOU have learned from the Report of the Council that the Society has considerably gained in number since the last Annual Meeting. So large an accession of members shows the growing popularity of our science, and is at once a gratifying reward of your past exertions and a sure presage of your further success. You have also been informed that during the same period the losses of the Society have been unusually numerous. Several of the deceased, whose main objects in life, if not alien, were connected but remotely with those of our institution, conferred upon it, notwithstanding, by their enlightened encouragement, important advantage: but the merits of the poet, the historian, the statesman, the warrior, though recorded in the annals of a grateful country, must not here be dwelt upon. To the memory of those only who have been closely allied to us, as fellow-labourers, will you desire that I should pay, individually, the well-earned tribute of our common regret.

The late Dr. Babington, whom we have been accustomed to look to with a respect almost filial, attached himself in early life to the study of chemistry and mineralogy. In the year 1795, he published a Systematic Arrangement of his collection of minerals purchased of the Earl of Bute, the finest, perhaps, which at that period existed in England; and in 1799, his New System of Mineralogy, which may be considered a continuation of the former work. These works, now superseded by others, which the introduction of improved modes of inquiry and the application of new instruments have rendered more perfect, evince much patient research and an exact knowledge of the state of mineralogy at that time. Active in the cultivation of science himself, Dr. Babington was quick to discern and eager to encourage merit in others. With a view to enable Count Bournon, of whom he had been a pupil, to publish his elaborate monograph on carbonate of lime, Dr. Babington, in 1807, invited to his house a number of gentlemen the most distinguished for their zeal in the prosecution of mineralogical knowledge. A subscription was opened and the necessary sum readily collected. This object having been accomplished, other meetings of the same gentlemen took place for the joint purpose of friendly intercourse and mutual instruction. From such small beginnings sprang the Geological Society; and among the names of those by whose care and watchfulness it was supported during the early and most perilous crisis of its history, that of Dr. Babington must always stand conspicuous.

A 2

But while Dr. Babington employed his leisure in the study of chemistry and mineralogy, he gradually rose into eminence as a physician, and at last became occupied with the care of a numerous family, and subjected to all the labour and responsibility of extensive medical practice. During many years, he was disabled from pursuing his favourite sciences with that unremitted attention which alone leads to original discovery; and accordingly our Transactions do not contain any communication from his pen: no man, however, more steadily cheered us in our progress or more heartily rejoiced in our success. In the year 1822, he was elected to the presidency of this Society, an office which he accepted in deference to the earnest wish of the Members, and held for two years at great personal sacrifice. His conduct in this chair afforded to us ample opportunity of observing the native goodness and kindliness of his heart, the urbanity of his manners, the evenness and cheerfulness of his temper, and the aptitude with which he exercised every liberal feeling.

During the presidency of Dr. Babington, and at his suggestion, was established the practice of submitting to immediate discussion the papers read at the table of the Society. Apprehensions were entertained by some persons at that time, that the collision of argument and the desire of personal distinction might interfere with the love of science or break the bonds of social intercourse,—that we might learn to contend less for truth than for victory. I appeal to you, Gentlemen, whether the brighter anticipations of Dr. Babington have not been amply justified by experience; whether our discussions, continued now during twelve years, have not been strongly characterized by a love of truth; whether the bonds of friendship have not been more closely cemented by them. Our conversations have been animated, but never intemperate; they have encouraged the timid, assisted the investigator in discovering the object of his research, and given additional value to every paper in our Transactions.

Dr. Babington was a Vice-President during the years 1810, 1811, 1812, 1813 and 1814, and a Trustee from 1811 to 1821. His donations to our library and museum were extensive, and from subscriptions set on foot to promote the objects of the Society his name was never withheld.

Dr. Babington retained to the latest period of his life a keen relish for the attainment of knowledge, and made considerable sacrifices to enable himself to keep up with its rapid progress. After descending from this chair he took private lessons in geology of Mr. Webster. So late as the winter of 1832–3 he enrolled his name at the University of London as a student of chemistry, and there attended with the utmost punctuality a course on that science of seven months' duration; he afterwards in the same spirit, and in his 77th year, once more applied himself seriously to geology, and went over the collection of fossils in our museum. I can scarcely imagine a more gratifying spectacle than that of a veteran in the labours of professional duty, thus returning to the pursuits which he had loved when young, and seeking relaxation, not in ease and repose, the allowable luxuries of old age, but in the indulgence of an enlightened passion for knowledge.

I need not apologize for these extended comments; they are more than justified by the occasion. The duties which your benefactor owed to the Society he cheerfully and fully performed. May his memory kindle in us a feeling not merely of gratitude but of emulation!

Dr. Berger, who died in the early part of last year, was a native of Switzerland, and had been employed in geological study for some years previous to 1813, when he sought in England an asylum from the foreign oppression which in those days of revolution had visited his country. In 1816, at the request of some of his friends in this Society, he agreed to devote himself for three years to geological investigations in the British Islands; and an annual sum was insured to him during that period by a subscription of some of our members. The north-west coast of Ireland was suggested for his first examination, and there, as might perhaps have been foreseen, the movements of a foreigner, who spoke our language imperfectly, and whose occupation must have appeared to the inhabitants mysterious, if not dangerous, at first excited doubt and obstruction, which, though not unamusing, were attended with some embarrassment, and called for the interference of his friends. He laboured with great zeal and assiduity, in that interesting field of inquiry, till his health unfortunately gave way His papers and collections were therefore incomplete ; and his attention appears to have been given perhaps too much to the investigation of details not immediately connected with the proper and immediate business of the geologist. His merit, however, must be judged of, not by reference to the present state of knowledge and the methods of inquiry now pursued, but to the condition of the science at that time. The facts he accumulated were valuable. " A Memoir on the Dykes of the north-east coast of Ireland," by himself, appears in the third volume of our Transactions ; his remaining papers were put into the hands of the Rev. William Conybeare, who subsequently went over the same country with Dr. Buckland; and we are indebted to the labours of Dr. Berger, extended and illustrated by these geologists, for one of the most valuable memoirs in the earlier volumes of our Transactions. The late years of Dr. Berger's life were passed in his native country, in bad health: he died at Geneva in 1833.

In perusing at the distance of so many years the record of the arrangement by which Dr. Berger's services were obtained for this Society, and the names subjoined*, I have been much struck by the delicacy with which his personal feelings were consulted, and have looked back with pride and exultation to the early history of our institution. I cannot be surprised at the success which has attended your exertions, when I call to mind the noble and disinterested spirit

* The paper bears, with the names of other Members who still remain, the signatures of the late Dr. Babington, Dr. Marcet, Mr. Francis Horner, Mr. Morgan, Dr. Wollaston, Sir Joseph Banks and Mr. Ricardo.

by which the first steps in your progress were directed. On no occasion since I have known the Geological Society, (and I have known it from its birth until the present hour,) have the Members hesitated to contribute, with the most liberal devotion, both personal labour and pecuniary support, whenever the *probable* advancement of science appeared to call for them. I mention this with double satisfaction, because I am convinced that this good spirit still subsists amongst us with undiminished vigour.

Dr. Alexander Turnbull Christie imbibed in the class-room of Professor Jameson a taste for geology, which he afterwards improved in India, as far as opportunity allowed, under many discouraging circumstances. On his return to Europe he applied himself to the science with great earnestness; he studied the best works, courted the society of their authors, familiarized himself with the contents of collections, and practised in the open air the most approved methods of investigation. He became the pupil of M. Brongniart at Paris, and the companion of M. de Beaumont and M. von Buch in the Alps. His studies were by no means confined to geology ; they embraced every department of natural history. The climatological and geographical distribution of plants was a subject to which he paid much attention. Having provided at his own expense the best instruments for the purpose, he returned to India with the design of instituting there a continued series of barometric, hygrometric, and other experiments, as well as of exploring the physical structure of that vast region, and of determining the relations of its rocks to those of Europe. On his way he visited Sicily, and transmitted to the Society an account of some of the younger deposits of that island, and the phænomena that accompanied their elevation. He wrote also a description of some bone caves near Palermo, and of tidal and other zones observed on limestone along the shores of Greece. These notices will be found in Jameson's Journal. Dr. Christie died prematurely of a jungle fever, while crossing the Nilgherry hill in November 1832.

Mr. Lansdown Guilding, though not himself engaged in the pursuit of geology, added several valuable specimens to our collection, and materially assisted the progress of some other branches of natural history, especially in connexion with the West Indies.

Sir Charles Giesecke was born at Augsburg in 1761. He was originally intended for the church ; after various changes of occupation and a life of some adventure, he devoted himself in about his fortieth year to mineralogy, and studied under Werner at Freyberg in 1801. He subsequently travelled with mineralogical views in several parts of the North of Europe ; in 1806 he entered into the service of Denmark and repaired to Greenland, leaving at Copenhagen a valuable collection of books and minerals, which were destroyed during the bombardment of that city. In Greenland he formed acquisitions of great interest in various departments of na-

7

tural history, but foreseeing the probability of their capture on the passage to Europe, he with great resolution and perseverance went a second time over the ground he had examined, and remained in that desolate region till his object was accomplished. In the mean time the vessel which contained his first treasures was taken, and the cargo sold by auction at Leith. The minerals attracted but little general notice, in part, I have been informed, from their being packed in moss and sea-weed, and perhaps also from the very circumstance of many of the species being unknown. Mr. Allan purchased nearly the whole collection, which upon examination proved to contain a great number of new and rare substances of the highest mineralogical interest, cryolite, sodalite, allanite, with mixed groups of striking variety and novelty; and all in such abundance that most of the cabinets of England (when collectors, if not more numerous, were at least more active than I fear they are at present,) were supplied from this source. Mr. Giesecké himself accidentally arrived at Leith in 1813, not long after Mr. Allan had published an account of his purchases, and with great generosity contributed to the improved catalogues and descriptions of specimens which subsequently appeared. He was soon after appointed Professor of Mineralogy to the Royal Dublin Society, and went to reside in Ireland, where he spent the remainder of his life. About this period also he was honoured with an Order of Knighthood by the King of Denmark; but having now passed his fiftieth year, his health was broken, and much of the energy lost which distinguished his early life. He lived to the age of 72, and died at Dublin in March 1833. Sir Charles Giesecké meditated, after his return from Greenland, an extensive work upon that country; he published a brief account of it in Dr. Brewster's Encyclopædia, but the larger work was deferred till the voyages of Ross and Parry had deprived the subject of the interest of novelty. His meteorological observations appeared in the Edinburgh Philosophical Journal for 1818; and he gave to Mr. Scoresby, for his work on the Greenland Coast, the use of his maps and other materials. The Edinburgh Philosophical Journal for 1822, contains an account of his discovery of the geological situation of Cryolite. His only publications on the mineralogy of Ireland are, I believe, a brief notice of the geological situation of Beryl in the county of Down *, and an account of an excursion to the counties of Galway and Mayo†.

Mr. Alexander Nimmo was a civil engineer of high reputation. He was born in Fifeshire in 1783, and at a very early age showed a strong propensity to physical and mathematical inquiry. One of his first public employments was a survey of some of the bogs in

* Annals of Philosophy, 1825. New Series, vol. x. pp. 74 & 75; republished from the Dublin Philosophical Journal.
† Annals of Philosophy, 1826; republished from the Dublin Philosophical Journal.

Ireland, on which he delivered a report to the Commissioners in 1811, containing some general observations on the geological character of part of Roscommon, Kerry, Cork and Galway. He was afterwards engaged in various works of great importance, principally in Ireland. He was the author of several articles in Brewster's Edinburgh Encyclopædia, on subjects connected with his profession. One of his latest and most valuable literary productions, on the publication of part of which he was engaged at the period of his death in January 1832, was a Chart of the Irish Channel, with sailing directions for the coast of Ireland, a performance probably connected with a paper which he laid before the Royal Irish Academy " On Geology as applicable to the Purposes of Navigation."

Mr. David Scott was one of the numerous class of officers in the service of the East India Company who have found means to combine with the most exemplary discharge of their official duties, a constant attention to the interests of literature and science. He was the second son of Archibald Scott, Esq., of Montrose, and died prematurely in India in 1831, at the age of 45, having passed through many offices of high trust with distinguished credit, and holding at the time of his death the situations of Civil Commissioner in Bungpoor and other districts, and agent to the Governor General in the North-east of Bengal. His exertions and success in discharging his official functions, and in promoting the welfare of the country in which he was placed, by diffusing education, were highly appreciated, and a monument has been erected to his memory by the Supreme Government of India. Mr. Scott possessed great knowledge in several branches of science not immediately connected with this institution, and lost no opportunity of attending to geological research. Our Transactions are indebted to him for the substance of a valuable paper communicated by Mr. Colebrocke*, " On the Geology of the North-eastern Border of Bengal," in which is described a remarkable deposit on the left bank of the Burrampooter river, containing an assemblage of fossils that bear an extraordinary likeness to those of the London clay. " Among the remains of fishes," Mr. Colebrooke states, " bony palates and the fins of the Balistes are common to the Indian clay and to that of Sheppey ; and the shells of Cooch-behar bear a strong generic, if not specific, resemblance to the marine formations above the chalk in France and England." This communication contains also some valuable facts respecting a succession of strata, like those of our coal-fields, in the Tista and Subuk rivers ; and in the same volume, is an extract from a letter written by Mr. Scott, describing the situation of a limestone and clay containing Nummulites at Robagiri, a village in the North-east of Bengal. Such resemblances, though they are far from establishing the contemporaneous formation—much less the continuity —of the groups in which they occur, are interesting, from the proof they furnish, of the operation of similar causes in very distant parts of the former surface of the globe.

* Series II. vol. i. pp. 132—140.

On the accounts of the past year put into your hands today, I will make but one observation. From the report of the Auditors it appears that the balance of disposable property in favour of the Society, taken at a very moderate estimate, is £2010, while the total amount of the compositions of all the compounders in the List of Fellows since the foundation of the Society does not exceed £2394. The difference is less than £400. If, then, the value of the collections, library, and furniture belonging to the Society be taken into account, our actual property considerably exceeds the claims of all our compounders, our current income being wholly disposable and free.

WOLLASTON MEDAL.—The product of the Wollaston Fund during the past year has been awarded to Mr. Agassiz of Neufchatel, in promotion of his work on the " General History of Fossil Fishes."
The first part of Mr. Agassiz's publication has but recently reached England, and the Council have availed themselves of the earliest opportunity of giving support to an undertaking of great geological importance. The author's qualifications for this work were so highly appreciated by the late Baron Cuvier, who had himself been engaged in a similar project, that on seeing Mr. Agassiz's collection of drawings, and hearing a statement of his views, and the results at which he had arrived, that profound naturalist at once transferred to Mr. Agassiz the whole of his materials. The approval of Cuvier is fully sanctioned by the portion of the work which is now before the Society. In deciding on the present award, the Council have acted strictly in compliance with the bequest of Dr. Wollaston. The work of Mr. Agassiz is intimately connected with the objects of this Society ; it demands for its completion great labour and expense. It is still in progress, and its publication has been ably commenced with a full assurance of the author's competency to the fulfilment of the task he has begun.
In his prospectus, Mr. Agassiz solicits the contribution of specimens from all quarters ; and I cannot better close the announcement of a testimony of approbation which l trust will be gratifying to his feelings, than by requesting the Fellows of the Geological Society to aid the progress of this important work, by giving or lending to its author any drawings and specimens of fossil fishes which they may either possess or obtain. The transmission and return of these loans can be easily effected through the medium of the officers of this house.

The History of Geology has been recently treated by several authors, especially by Mr. Conybeare and Mr. De la Beche, in a manner which would render any observation from me on that subject at once superfluous and imprudent. The communications read at our general meetings have been fixed in your memory by the discussions to which they have given rise, and the published abstract of their contents. Still, however, it may be well to enumerate these communications, that you may measure the exertions made here since the last Anniversary, and the effect they have had on the state of geological knowledge.

MISCELLANEOUS.

The experiments of Sir James Hall mark an important epoch in science. It was with great delight, therefore, that we received from Captain Basil Hall, R.N., a collection of the products of these experiments, and some of the instruments with which they were conducted. Among the latter is a machine for regulating high temperatures, accompanied by an account of its properties and mode of acting.

Mr. Gardner, the well known geographer, has drawn our attention to the curious fact, that of the land on the surface of the globe only $\frac{1}{27}$th part has land at its antipodes.

Sir David Brewster has communicated to us his interesting observations on the properties of the diamond, from which it would appear to be of vegetable origin,—the cavities whence these properties are derived being found in amber, but not in any product either of igneous fusion or of aqueous solution.

HOME GEOLOGY.

Dr. Mitchell has laid before the Society a detailed account of the geology of Harwich in Essex, of the Reculvers in Kent, of Quainton and Brill in Buckinghamshire. Mr. Dadd has described the Vale of Medway and its neighbourhood. Dr. Fitton, who published in the early part of the year, a geological sketch of the vicinity of Hastings, has supplied us with an account of some instructive sections recently exposed to view at St. Leonard's. Mr. Woodbine Parish has sent to us portions of the Iguanodon and Lepidosteus from the well known " White Rock," situate in the same district, and now almost destroyed. Our knowledge of the inland Extent of the Wealden Formation has been enlarged by a paper of Mr. Strickland, accompanied by specimens of Paludina from the ferruginous sand of Shotover hill.

Mr. Strickland has also rectified the boundaries of some of the strata near Bewdley, and traced a line of fault from the north of Bredon Hill to Little Inkbarrow.

Sir Philip Egerton has supplied us with further information in respect to the lower portion of the Connaught Coal-district. Beneath the coal at Kulkeagh in the county of Fermanagh is a shale 600 feet thick, with subordinate layers of black marlstone and clay-iron ore towards the top, and a thin stratum of micaceous grit near the bottom. All the beds are replete with ammonites, orthocera, producta, encrini, corals and calamites. This deposit lies on sandstone separated by the mountain limestone from another bed of shale marked by characteristic fossils, and the entire system therefore appears to bear a strong resemblance to the lower portion of the carboniferous beds in the South-west of England.

In the carboniferous strata of Coalbrookdale, Mr. Prestwich has described a heterogeneous assemblage of plants and shells both of fresh- and salt-water species. A band of ironstone, nearly in the centre of this series, contains four genera of Trilobites: in the same coal-field Mr. Anstice has recognised two genera of insects. On the opposite side of the Severn, Mr. Murchison has found at Pontesbury, Uffington, Le Botwood and other places, a band of compact limestone, between two beds of coal, resembling the lacustrine limestone of central France, and containing freshwater shells. These discoveries may throw light on those which have been since made at Burdie-house and elsewhere in the neighbourhood of Edinburgh.

The structure of other coal-fields has been illustrated by Mr. Murchison, Mr. R. J. Wright, and Mr. England.

After careful examination of the Old red Sandstone, Mr. Murchison has proposed to divide it into three parts : the uppermost, characterized by quartzose Conglomerate ; the middle, by Cornstone ; the lowermost, by Flagstone. The cornstone and marlstone of the middle group contain undescribed genera of crustacea ; and in the tilestone beneath are found some defences of fish, together with a few remains of testacea.

Mr. Murchison has employed three summers in examining a range of country situate between Shrewsbury and Caermarthen; and the geological positions as well as the mineral and zoological characters of the several rocks which border England and Wales are now determined with as much exactness as those of any portion of the secondary system. Taking the old red sandstone as a line of departure, the rocks beneath are disposed in descending order as follows:

1. The Ludlow series, divisible into three parts, the upper, middle and lower. To the middle belong the well-known limestones of Amestry and Sedgley : the upper and lower consist of sand, marl, or flagstone, having some fossils peculiar to each, and others in common. The thickness of the whole is estimated at 1000 feet.

2. The Dudley or Wenlock series, consisting of limestone : its thickness may be taken at 2000 feet.

3. The Hordesley or May Hill series, composed of party-coloured sandstone, conglomerate and impure calcareous flagstone : it is said to attain a thickness of 2500 feet.

4. The Built or Llandilo series, a black flagstone, characterized by the *Asaphus Buchii.*

5. The Longmynd or Linley series, consisting of coarse roof slate, sandstone and conglomerate ; no fossils have been discovered in it.

It is well known that Professor Sedgwick has studied with equal assiduity the rocks which lie beneath those I have mentioned. When his observations are published, the Society will have a type of the whole of the transition rocks of Wales. The rocks described by Mr. Murchison are, for the most part, exceedingly well characterized by their fossil contents. Some of the shells which he has discovered, appear to have escaped the notice of antecedent observers ; but the genera, if not the species, of others, may occasionally be found in the works of Hisinger and other continental writers. If, then, the transition as well as the secondary and tertiary beds can be identified over great tracts of country by their fossil remains, let us hope that a clue is now at hand, by which we may find our way through that vast assemblage of beds, which, not in England only, but in Scotland, Ireland, Germany, Russia, Sweden, and North America, has hitherto presented to the observer a mere scene of confusion. In Mr. Murchison's paper we find also, traced with exactness, several hitherto unexplored lines of disturbance, producing sometimes, as in the Abberley Hills, a complete inversion of dip. The rocks which border the old red sandstone, acquire in some places an anticlinal dip, and reappear in parallel ridges far westward of their natural site, insomuch that the Ludlow series is met with even in Montgomeryshire. Mr. Murchison has examined in detail the trappean and porphyritic rocks to which these disturbances are for the most part assignable, but the description of them has been reserved for communications not yet before us.

Professor Sedgwick has transmitted to us a notice on the granite of Shap in Westmoreland. From recent excavations it appears that veins of this granite penetrate the adjoining strata, from which he infers that it is of posterior date.

Mr. De la Beche, one of our Vice-Presidents, acting under the direction of the Board of Ordnance, has produced a geological map of the county of Devon, which, for extent and minuteness of information and beauty of execution, has a very high claim to regard. Let us rejoice in the complete success which has attended this first attempt of that honourable Board to exalt the character of English topography by rendering it at once more scientific and very much more useful to the country at large.

Organic Remains.—Every succeeding year brings to light new fossil animals which cannot be assigned to existing genera. Dr. Riley, deeply skilled in physiology and comparative anatomy, has given us an account of an animal so extraordinary, that naturalists differ even respecting its class. After careful examination, he considers it a cartilaginous fish, partaking of the character both of the Rays and the Squales. Here then is another instance of a link, now wanting to connect existing genera, having formerly existed.

Towards the close of the last session Mr. Channing Pearce exhibited to the Society a matchless collection of Apiocrinites found at Bradford in Wiltshire. To the description of this fossil as given by the late Mr. Miller, Mr. Pearce adds that the column was occasionally ten inches long. He has found in the great oolite, three species of Apiocrinites, differing in the form of their body, and the thickness of its component plates.

Foreign Geology.

Europe.—The structure of the South of Spain has been illustrated by Colonel Silvertop and Captain Cook. From the joint labours of these gentlemen we learn, that the country between the Sierra Morena and the Mediterranean consists of lofty ranges of granite, slate, serpentine and limestone, succeeded either by red sandstone or by vast beds of secondary, compact, dolomitic limestone.

13

We also learn from them that the valleys and plains which border the shore of the Mediterranean, are composed of tertiary strata; but we are indebted solely to Col. Silvertop for pointing out to us, on the authority of M. Deshayes, that the tertiary deposit of Malaga and the districts adjacent belongs to the Pliocene, while that of the basins of Baza and Alhama belongs to the Miocene epoch.

Mr. Lyell has laid before us an account of the lignite formation of Cerdagne in the Eastern Pyrenees. This lacustrine deposit reposes in horizontal beds on granite and hornblende and argillaceous schist at the height of 3000 and 4000 feet above the level of the sea. The shells procured are too imperfect to determine its age.

A memoir on the neighbourhood of Bonn was presented last year by Mr. Horner. After describing the characters of the grauwacke, trachyte, basalt, brown coal, gravel and löss, the author compares the age of these with that of analogous formations in other parts of Europe, and of one another. The beds of grauwacke as they contain Terebratulæ and other shells he refers to the upper part of that system; he considers the brown coal more recent than the plastic clay, some of its plants and shells having been identified with specimens found at Aix en Provence. The löss, which reposes on a thick bed of gravel, and contains existing land shells, together with bones of extinct quadrupeds, is considered the latest deposit, and attributed to the bursting of a lake in the upper part of the Rhine. From the beds of trachytic tuff being interstratified with brown coal, and from the occurrence of a bed of basalt above it, Mr. Horner infers that volcanic operations took place during, and even subsequently to, the deposition of the lignite. Having thus established the comparative age of the brown coal, he also determines that of the volcanic rocks.

The tertiary coal or lignite near Gratz, in Styria, is interesting on account of its organic remains. In the memoir of Professor Sedgwick and Mr. Murchison on the Eastern Alps, the strata of this deposit, which are nearly horizontal, are shown to rest on " an inclined system of secondary green-sand." Imbedded in the coal are various vegetable remains, shells of a Cypris, scales of fishes, and fragments of bones of Mammalia and Tortoises. Professor Anker of the Joanneum, has sent to the Society an account of these, together with the drawing of a jaw, which Mr. Clift conceives to have belonged to a Hyæna.

Mr. Pratt, ignorant of the prior researches of Dr. Christie, carefully examined, in the year 1832, the caves of Monte Grifoni near Palermo; and having ascertained the height to which the perforations of lithodomi extend in each, infers that the change of level was not effected by one movement, but by several.

ASIA.—Much information has been received from the East during the past year. Mr. Burnes, distinguished for his travels in India, Persia and Toorkistan, has presented to the Society his geological memoranda of the countries lying between the mouth of the Indus and the Caspian Sea. Mr. Burnes, though he did not travel for the express purpose of studying geology, carefully and faithfully noted whatever attracted his attention. In reading his account of these hitherto almost unknown regions, we cannot but be struck with the resemblance of their geological structure to that of Europe. The central axis of the Hindoo Koosh is composed of granitic rocks, succeeded by various schists, conglomerates, variegated marls, limestones and sandstones. Besides this mighty system, some portion of which cannot be identified with European strata for want of fossils, there is a vast range of salt (previously noticed by Mr. Elphinstone), of coal, and, near the mouth of the Indus, nummulitic limestone.

In a late number of Jameson's Journal is part of a memoir on the structure of the Valley of Ovelipore* by Mr. Hardie, one of our recently elected Fellows. This valley had previously been noticed by Captain Dangerfield†; but Mr. Hardie has been the first to describe a singular Indian formation which occurs there, called Kunkur. It is rarely, if ever, stratified; it forms a bed, seldom exceeding a few feet thick, which mantles over the irregularities of the country. It is sometimes imperfectly oolitic; at others globular, botryoidal or nodular; in some places a compact limestone; in others it resembles chalk : not unfrequently it contains round and angular fragments of rocks. No animal or vegetable remains have been noticed in it. The author carefully distinguishes Kunkur from modern tufaceous deposits, but assigns to it a similar origin.

AMERICA.—Captain Colquhoun and Mr. Burkart have presented to us a specimen of native iron from Zacatecas, and memoranda on this and similar masses found in Mexico.

Captain Bayfield has communicated to us a paper on the shores of the River and Gulf of St. Lawrence from the Saguenay to Cape Whittle. The information contained in this memoir completes our knowledge of the north coast of the St. Lawrence‡; and from the previous labours of Mr. Green in the district of Montmorency§; of Lieutenant Ingall in the country bordering the rivers St. Maurice and aux Lievres‖; of Captain Bonnycastle in Upper Canada¶; of Dr. Bigsby**, Captain Bayfield†† and Dr. Richardson‡‡, on the shores of Lakes Ontario, Erie, Huron and Superior; and of Dr. Richardson in the overland expeditions to the Arctic Seas, we have a general account of the geological structure of the whole country between the mouths of the Mackenzie and Copper Mine rivers and the

* The city of Ovelipore is in lat. 24° 25′ N. long. 73° 44′ E.
† See Sir John Malcolm's Central India.
‡ See on the country between the St. Maurice and the Saguenay, Trans. Quebec Society, vol. ii. p. 216. On the Saguenay country and St. Paul's Bay, ibid. vol. i. p. 79; vol. ii. p. 76. On Quebec, Proceedings Geol. Soc. No. 5, p. 37.
§ Quebec Trans. vol. i. p. 181.   ‖ Ibid. vol. ii. p. 7.
¶ Ibid. vol. i. p. 62.
** Proceedings Geol. Soc. No. 3, p. 23. Trans. Geol. Soc. Series II. vol. i. p. 175. Journal Royal Institution, vol. xviii. pp. 1, 228.
†† Quebec Trans. vol. i. p. 1.   ‡‡ Appendix, Expedition to Polar Seas.

Gulf of St. Lawrence. The researches made during the expeditions
of Captain Ross, Sir Edward Parry and Sir John Franklin, have
also given us a general insight into the nature of the formations
which constitute a large portion of the shores of the Western
Polar Seas. Why should I repress the feeling of patriotic pride
which rises within me on contemplating how vast a range of the
western continent has thus, in the brief period of a few years, been
brought within the pale of our science almost entirely by the exertions of English officers? Great is the gratitude we owe them;
yet have their services not been wholly without reward. The
taste for scientific research which sprung up in the minds of these
gallant men, spontaneously, as it were, and without the aid of regular systematic culture, has been to many of them a welcome relief
from the toil and monotony of professional duty; while to others it
afforded pleasurable occupation in the solitude of trackless deserts,
under exposure to all the rigour of an arctic climate, in the absence
of European indulgences, and even under the terrible apprehension
of impending starvation.

The district surveyed by Captain Bayfield is bounded by hills,
composed of granite, sienite and trap rocks, which enter so largely
into the structure of the two Canadas. Clay, sand and gravel,
apparently recent, occupy the coast. The Mingan, the Esquimaux
and Anticosti Islands are of limestone, containing fossils like those
of Lake Huron. But the most interesting feature in this communication is the evidence it affords of a change in the relative
position of land and water. In the Mingan Islands is a series
of shingle terraces, agreeing in character with the recent beach,
the most distant being 60 feet above the level of the highest
tide. The author describes, with great care, the different vegetation of each terrace, the one furthest from the shore being covered
with trees, the nearest almost barren; parallel to the shore, in this
island, natural columns of limestone have been scooped out by the
action of water at different periods; the levels of the water-worn portions agree with those of the terraces, and the depth of the scooped
parts, with the rise of the present tidal wave of the St. Lawrence.
Captain Bayfield has noticed similar terraces on the adjacent main-
land and in the neighbourhood of Quebec, and thinks the phænomena indicate successive elevations of the land rather than successive depressions of the water.

Among the subjects which have for some years past engaged the
thoughts of geologists, none perhaps has excited so general and intense an interest as the Theory of Elevation. I shall avail myself,
therefore, of the present occasion to lay before you a connected
statement of the scattered facts and opinions upon which it rests.

On entering upon this subject, it is necessary to understand distinctly what is meant by Elevation. Definitions have recently
been decried, I think unwisely. The formation of definitions,
it has been said, and the establishment of unerring distinctions
are among the last, and not the first steps of systematic knowledge. Equally true, and far more salutary is the lesson that sci-

16

ence cannot be advanced by equivocation. As in trading concerns fixed weights and measures are necessary guards against fraud, so in philosophical investigation words of definite meaning are indispensable securities against sophistry and self-delusion. Euclid did not end, he began with defining. Mathematical certainty has no other basis than mathematical precision, and the greater part of those absurdities which from time to time attach themselves to all other branches of knowledge derive their subsistence from ambiguity of language and a dearth of definition.

A torrent brings down a quantity of alluvial matter, and the plain on which it rests is said to be *elevated*.

An opening occurs in the earth; ejected ashes, scoriæ and lava accumulate around it; a Monte Nuovo is formed; and the areait occupies is said to be *elevated*.

By the persevering labour of polypi, a coral reef gradually attains the surface of the ocean; and the fabric so constructed is said to be *elevated*.

A porous rock covers a rock that is not porous; the rain filters through the superincumbent bed; springs break out in the subjacent; and at last, for want of support, the porous rock, originally horizontal, acquires an inclined posture, one end being directed upwards, the other downwards; and the whole is said to be *elevated*.

An earthquake takes place at the mouth of a river; the sea is violently affected; a bar is formed at the entrance of a harbour from the washing in of new alluvion, or from some obstruction to the escape of the old; where a ship floated, a barge is aground; and the land is said to be *elevated*.

Such instances of Elevation are common and incontestible; but elevation of this kind is quite different from that which forms the subject of my present inquiry.

By the term *Elevation*, I mean only the removal of any given object from a lower level to a higher level; consequently it is necessary, before I speak of an object as *elevated*, that I should be prepared to show two things: first, the level at which it has stood; secondly, the level at which it stands.

That I might form a right opinion of the theory, the merits of which I am about to investigate, I have endeavoured to determine the site, the number and the magnitude of those multifarious objects to which the attribute of elevation is continually applied. The attempt has proved unsuccessful: they are indefinite in place, in form, and in dimension. That Mountains should be elevated is not surprising, but we are familiarized also with Valleys of elevation*. In ancient ti mes an Island (Delos, for example,) would alternately

* Valleys of this nature are properly called by Mr. Scrope "valleys of "elevation and subsidence," or more concisely, "anticlinal valleys." See Scrope on Volcanoes, p. 213.

emerge from, and plunge beneath, the sea. Extensive Provinces, nay, entire Kingdoms, now perform the same feat. The existence of Craters of Elevation is by some still considered doubtful; but it is an accredited fact that Mountains and Mountain Chains have risen, either *per saltum* or *per gradus.* All the Strata have been raised; and all Unstratified Rocks would doubtless have been raised also, but that some have risen of themselves. The Bed of the Sea has been elevated again and again. Continents too have been raised, though " by an operation distinct from that which raised the Pri- " mary Strata."

The arguments advanced in favour of these doctrines are derived either from observation, or from induction.

It is stated by Von Hoff, that in the year 1771 several tracts of land were upraised in Java, and that a new bank made its appearance opposite the mouth of the river Batavia. The authorities cited for the effect of this and several other earthquakes mentioned in the same place by this author, are Sir Stamford Raffles, John Prior's Voyage in the Indian Seas, and Hist. Gen. des Voy. tom. ii. p.401. Mr. Lyell has cited the first of these only, but no such fact is noted in either edition of the work of Sir Stamford Raffles. The other authorities adduced by Von Hoff I have been unable to consult; but from the Appendix to the Batavian Transactions (which contains an apparently authentic account of all the recorded earthquakes that have taken place in Java during a century and a half,) it would seem, that in the year 1771, in which the uprising is said to have happened in that island, there was no earthquake at all.

The Earthquake of Chili in 1822 has been so much* insisted on, that it requires detailed consideration. Of this event an account by Mrs. Graham is inserted in our Transactions. I am deeply sensible of the honour that lady conferred on the Society by her obliging compliance with the request which elicited her narrative, and it is only the importance of its contents which could induce me to subject them to the test of rigid examination.

According to this account " it appeared on the morning after the " earthquake, that the whole line of coast from north to south, to the " distance of above 100 miles, had been *raised* above its former level." But by what standard was the former level ascertained? who on the morrow of so fearful a catastrophe could command sufficient leisure and calmness to determine and compute a series of changes, which extended 100 miles in length, and embraced (according to a statement in the Journal of Science,) an estimated area of 100,000 square miles? How could a range of country so extensive be surveyed while the ground was still rocking, which it continued to do on that day, and for several successive months? What was the average number of observations per square mile? Who made, checked and registered them? By what means did the surveyors acquaint

* Bakewell's Geology, edition 4, pp. 98. 504. Lyell, vol. i. pp. 401. 455. De la Beche's Manual, edition 2. Scrope on Volcanoes, p. 209.

themselves with what had been the levels and contour before the catastrophe took place, by which, as we are told, all the landmarks were removed, and the soundings at sea completely changed?

Mrs. Graham states that by the dislodgement of snow from the mountains, and the consequent swelling of rivers and lakes, much detritus was brought to the coast ; and further, that sand and mud were brought up through cracks to the surface. Amid so many agents it should not be easy to assign to each, its share in the general result.

That fishes lay dead on the shore may prove only that there had been a storm. In her published travels, Mrs. Graham represents them as lying on the beach, which may very well have been thrown up, as the Chesil bank has been, by a violent sea. Some muscles, oysters, &c., still adhered, she says, to the rocks on which they grew ; but we know not the nature or dimensions of these rocks, whether fixed or drifted. The occurrence of a shelly beach above the actual sea level is an observation which must not be lost sight of. I propose to speak of it hereafter : in the mean time be it recollected, that these beaches are said to occur along the shore at *various* heights, along the summit of the highest hills, and even among the Andes.

Neither in the paper of Mrs. Graham, nor in the anonymous account published about the same time in the Journal of Science, can I find any paragraph to justify the position (which, from the seductive character of the work* in which it appears, may, if not now assailed, soon be deemed unassailable,) that a district in Chili, one hundred thousand miles in area, " was *uplifted* to the average height " of a foot or more ; and the cubic contents of the *Granitic Mass* " added in a few hours to the land." By what means we get the average I do not know. Mrs. Graham says the alteration of level at Valparaiso, was about three feet ; at Quintero, about four feet : but *the granitic Mass!* has the geological structure of Chili been sufficiently examined to assure us that Granite extends over one hundred thousand square miles ?

In the well-known work of Molina, a Jesuit who passed the greater part of his life in Chili, and wrote a natural history of that country, I find no ground for supposing that in any earthquakes which took place there from the time the Spaniards first landed on its shores to the date of his publication, any similar phenomena had been noticed. Moreover, the statement of Mrs. Graham, and of the writer before alluded to, respecting the *Elevation of land* which occurred during the earthquake of 1822, has not been confirmed by Captain King, nor by any naval officer or naturalist who has since visited that region, though many have visited it who had heard the circumstance, and who would willingly have corroborated it if they could. But they saw no traces of such an event ; and the natives with whom they conversed, neither recollected nor could be induced to believe it.

The 16th number of the " *Mercurio Chileno,*" a scientific Journal, contains an account of this earthquake, by Don Camilo Enriquez,

* Lyell, vol. i. p. 473.

which I have not been able to procure. A later number refers to this account, and to another published in the *Abeija Argentina*, a work of considerable reputation, which, by the kindness of Mr. Woodbine Parish, I have been enabled to consult. The account there given of the earthquake of 1822, is strongly recommended to the reader, " as a sensible straight-forward description of what actually took " place, without the high colouring in which ignorance and terror " and exaggeration are apt to indulge."

No notice is here taken of the permanent *Elevation of the Land,* and the account concludes thus:

" The earth certainly cracked in places that were sandy or " marshy; I saw cracks too in some of the hills, but mostly in " the low nook where much earth had run together; the sea was " not much altered,— it retired a little, but came back to its old " place. Don Onofri Bunster, who, on the night of the earthquake, " was walking on the shore at Valparaiso, in front of his house, had " a mind to go up on the hill, but could not, so great was the quan- " tity of falling dust and stones: he repaired to his boat therefore, " and with some difficulty got aboard; this done, he made obser- " vations on the motion of the sea; on sounding, the depth was " thirteen fathoms; he heaved the lead a second time, and the " depth was no more than eight fathoms: this alternate ebbing and " flowing lasted the whole night, *but did not the slightest harm on* " *shore.*"

These are the only cases I remember to have met with, in which the testimony of eye-witnesses has been adduced to prove the Rise of land by Earthquakes. That such Rise may have taken place, at different times, without being recorded, perhaps even without being observed, is not very improbable; but if I am to pronounce a verdict according to the evidence, I believe there is not as yet one well authenticated instance in any part of the world, of a non-volcanic Rock having been seen to rise above its natural level in consequence of an Earthquake.

Before I quit this subject, it may not be amiss to mention, that on comparing the times at which the successive shocks took place in Chili, as given by Mrs. Graham, and the other authorities to which I have had occasion to refer, the discrepancy is extraordinary.

I have already intimated in a few words, my opinion as to the sense in which land can be said *to be elevated by means of Volcanoes.* Of these, Vesuvius is perhaps the most constantly observed; and among the innumerable authors who have described its effects, from the time of Pliny down to the present day, not one pretends that the Apennine limestone, close at hand, has been in the least raised by that volcano. We shall do well to bear this in mind, when we have occasion to consider the height at which tertiary shells are found on Etna. That those shells belong to beds thrown up by Etna, is a doctrine founded upon induction, not upon experience. As far as experience goes, we have no reason to think that Etna, in its most violent paroxysms, will ever raise those tertiary strata above their present level.

Leaving these scenes of paroxysmal violence, let us next inquire, whether there may not be going on, in the calmest seasons and in the stillest countries, a *chronic and almost imperceptible impulsion of land upwards*.

As early as the time of Swedenborg, who wrote in 1715, it was observed that the level of the Baltic and German Ocean was on the decline. About the middle of the last century an animated and long-continued discussion took place in Sweden, first as to the cause of this phenomenon, and then as to its reality. Hellant, of Tornea, who had been assured of the fact by his father, an old boatman, and who afterwards witnessed it himself, bequeathed all he had to the Academy of Sciences, on condition that they should proceed with the investigation : the sum was small, but the bequest answered the purpose. Some of the members of the Academy made marks on exposed cliffs and in sheltered bays, recording the day on which the marks were made, and their then height above the water. The Baltic affords great facility to those who conduct such experiments, as there is no tide, nor any other circumstance to affect its level, except unequal pressure of the atmosphere on its surface and on that of the ocean : this produces a variation which is curiously exemplified at Lake Malar near Stockholm. As the barometer rises or falls, the Baltic will flow into the lake, or the lake into the Baltic. The variation resulting from the inequality of atmospheric pressure, however, is trifling. In sheltered spots, mosses and lichens grow down to the water's edge, and thus form a natural register of its level. Upon this line of vegetation marks were fixed, which now stand in many places two feet above the surface of the water.

In the year 1820–1, Bruncrona visited the old marks, measured the height of each above the line of vegetation, fixed new marks, and made a Report to the Academy. With this Report has been published an Appendix by Halestrom, containing an Account of Measurements made by himself and others along the coast of Bothnia. From these documents it would appear, 1. That along the whole Coast of the Baltic the water is lower in respect to the land than it used to be. 2. That the amount of variation is not uniform. Hence it follows, that either the Sea and Land have both undergone a *change of level*, or the Land only ; a change of level in the Sea only will not explain the phenomena.

A quarter of a century has now elapsed since Mr. von Buch declared his conviction that the surface of Sweden was slowly rising all the way from Frederickshall to Abo, and added that the Rise might probably extend into Russia. Of the truth of that doctrine the presumption is so strong, as to demand, that similar experiments and observations should be instituted and continued for a series of years in other countries, with a view to determine whether any change of level is slowly taking place in those also. The British Association for the Advancement of Science have already obeyed the call. A committee has been appointed to procure satisfactory data to determine this question as far as relates to the coasts of Great Britain and Ireland, and I cannot but hope that similar investigations will also be set on foot along the coasts of France and Italy, and eventually be extended to many of our colonial possessions.

The inductive arguments in favour of the *Elevation of land*, whatever the size, and whatever the amount of Rise, are founded chiefly on the following circumstances: 1. The height of sedimentary beds and marine bodies, whether corresponding or not to those of adjacent seas, or of the actual globe. 2. The height of terraces resembling sea beaches. 3. The height of ripple marks. 4. The change of posture which horizontal strata undergo in the neighbourhood of "unstratified rocks." 5. The various heights at which the same rocks occur in different parts of their course. 6. The anticlinical posture of strata frequent in, though not confined to, mountain chains. 7. The arched or domed configuration of some strata. 8. The occurrence of coral, apparently recent, high above the present surface of the sea. 9. The position of ancient buildings, viz. the temple of Serapis at Puzzoli, &c. I have not time to consider these arguments in detail; each deserves to form the subject of a separate treatise. Some of them prove not Elevation, but only change of level, which Subsidence would explain equally well. Some prove local disturbance, whereby one portion may have been thrown up, the other down. Some again afford a fair presumption of real *local* Elevation or Ascent. Most of them are good to a certain point: all are continually overstrained; and I am frequently astonished to observe how prodigious the weight, how slender the string that supports it.

The assigned *Causes of Elevation* are exceedingly various. One author raises the bottom of the sea by earthquakes; another, by subterranean fire; another, by aqueous vapour; another, by the contact of water with the metallic bases of the earth and alkalis. Heim ascribes it to gas; Playfair, to expansive force acting from beneath; Necker de Saussure connects it with magnetism; Wrede, with a slow continuous change in the position of the axis of the earth; Leslie figured to himself a stratum of concentrated atmospheric air under the ocean, to be applied, I suppose, to the same purpose.

It is impossible within the narrow limits of this discourse, that I can enter into the merits of these and other hypotheses seriatim. I must therefore throw them into two classes, the first of explosive forces, the second of sustaining forces; they are one and the same in Plutonic language, but still it will be convenient to separate them.

That explosive forces exist, or may exist, under the surface, no one can deny; but I cannot adopt the opinion (however high the authority from which it comes,) that " in volcanic eruptions we find a power " competent to raise *Continents* out of the ocean." The force we find in volcanic eruptions is limited in time, place and action ; it fuses bodies of easy fusibility; it tosses up those that are refractory, and thus forms either a current of lava or a shower of stones, scoriæ and ashes. What resemblance is there between this operation and the rise of a continent? With more propriety might it have been said that in a molehill we behold the action of a cause competent to raise mountains.

If by *Continent* is meant a whole Continent, and nothing but a Continent, its rise, provided this happened only once, would seem difficult to understand; but to me still more incomprehensible is the confident assurance we continually receive from writers of high and deserved reputation, that this event has happened again and again. Before we admit the Submersion of a continent, we must admit either that at a period immediately preceding that catastrophe, there existed under the land a cavity large enough to contain the continent about to be submerged, or that during the process the subjacent beds shrunk in consequence of a reduction of the temperature, and to such an extent that the contraction in a vertical line equalled the distance from the level of the highest tops of the continent to that of the surrounding ocean. In like manner, before we can admit the Elevation of a continent, we must admit either that, at a period immediately preceding that catastrophe, there happened an inroad of sustaining matter equal in thickness and in extent to the Continent about to be uplifted, or that during the process the subjacent beds expanded in consequence of an increase of temperature, and to such an extent that the expansion in a vertical line equalled the distance from the level of the highest tops of the continent to that of the surrounding ocean. These therefore are the events which we are taught to credit, as having taken place again and again, notwithstanding the tendency which caloric has to diffuse itself, and the apparently unaltered dimensions of the fissures and local caverns by which the strata are so often separated or intersected.

I will not expend more of your time in arguing against such doctrines. All men are more or less lovers of the marvellous, but few, I think, will upon reflection approve such marvels as these.

Solids, fluids and aeriform substances exist, we know, in the interior of the earth, and expand by heat, which exists there likewise. All of these, therefore, are fit *Agents of Elevation*, subject to certain conditions.

Dr. Daubeny attributes the liquefaction of lava, the throwing up of ashes, and all other phenomena of disturbance attendant on volcanic eruptions, to the Action of Water upon the Metallic Bases. This cause is not opposed to experience, and appears well proportioned to the effect, which is sudden, violent, occasional, temporary, accompanied by heat and by flame. To me, at least, it seems far more satisfactory than the explanation of those who ascribe the effect to the Elastic Power of Subterranean Fires, repressed in one place and relieved in another, or to the Undulations of a Heated Nucleus.

A heated *Central Nucleus* is a mere invention of fancy, traceable, I believe, to no other source than the hope of obtaining a good argument from the multiplication of bad ones. To the Huttonian and every other geological sectary who relies on this postulate, I say, be cautious; " *incedis per ignes dolosos.*"

The only observation I recollect to have met with in favour of central heat is, that the deepest mines are the warmest—be it so! Might not a geologist by parity of reasoning argue thus?—In travelling from Rome to Chamonix, the country becomes continually more and more mountainous ; some of the peaks of Chamonix are from ten to fifteen thousand feet above the level of the sea. Imagine, therefore, what they must be at Hamburgh!!

If mines derive their temperature from heat lodged in the centre of the earth, the temperature ought to vary with their distance from the centre, and therefore, since the earth is an oblate spheroid, the mines of Scandinavia ought at the same depth from the surface to be proportionally warmer than those of tropical countries; a result which has never been, I believe, even suspected.

The existence of *Central Heat* in the sense and to the extent assumed in the Huttonian theory, is contrary to all our experience. If Heat there be in the Centre of the globe, it must have the properties of heat and none other. I ask not how the Heat originally was lodged in that situation, for the origin of all things is obscure; but I ask why, in the countless succession of ages which the Huttonian requires, the Heat has not passed away by conduction, and if it has passed away, by what other heat it has been replaced?

Dr. Chalmers in speaking of Sir Isaac Newton, observes, that it was a " distinguishing and characteristic feature of his great mind, that it " kept a tenacious hold of every position which had proof to substan- " tiate it; but a more leading peculiarity was, that it put a most de- " termined exclusion on every position destitute of such proof. The " strength and soundness of Newton's philosophy was evinced as much " by his decision on those doctrines of science which he rejected, as " by his demonstration of those doctrines of science which he was " the first to propose. He expatiated in a lofty region, where he met " with much to solicit his fancy, and tempt him to devious speculation. " He might easily have found amusement in intellectual pictures, he " might easily have palmed loose and confident plausibilities of his " own on the world. But no, he kept by his demonstrations, his mea- " surements, and his proofs."

Gentlemen, let us, as far as is consistent with the nature of geological investigation, show the strength and soundness of our philosophy in the same manner.

That Heat of considerable intensity prevails occasionally, in certain places, at some depth, is all that we have as yet clearly established. Whether that Heat is permanent, whether it is generally diffused, whether it is central, are questions of mere speculation.

Intimately connected with the hypothesis of *Central Heat* is that of *Refrigeration.*

It has been observed by one of our members, that "the Remains both " of the animal and vegetable kingdom preserved in strata of different " ages, indicate that there has been a great Diminution of Tempera- " ture throughout the northern hemisphere, in the latitudes now oc- " cupied by Europe, Asia and America ; the change has extended to " the arctic circle as well as to the temperate zone; the heat and " humidity of the air, and the uniformity of climate, appear to have " been most remarkable when the oldest strata hitherto discovered " were formed. The approximation of a climate similar to that now " enjoyed in these latitudes, does not commence till the æra of the

" formations termed tertiary ; and while the different tertiary rocks
" were deposited in succession, the Temperature seems to have been
" still further lowered, and to have continued to diminish gradually
" even after the appearance of a great portion of existing species upon
" the earth." The little knowledge we have of the fossil productions
of countries south of the temperate zone, induces me to believe that
these observations are as applicable to the southern hemisphere as to
the northern.

This *Refrigeration*, one of the most undoubted facts in geology, is
supposed by the Huttonians, and if I mistake not, by M. Elie de Beaumont and others, to arise from a decrease of the *Central Heat*; an
opinion, however, which cannot, I think, be supported.

We know of one method only by which Central Heat, if it exists, can
pass from the earth, viz. by Radiation. It cannot pass by Conduction.
Conduction implies conductors, which in empty space are not to be
procured *, but the Radiation of heat, at low temperatures, is so slight
that it is scarcely sensible at 100° of Fahrenheit's thermometer,
a temperature twice as great as the medium temperature of the surface of the globe at this time. The Temperature of the earth's surface
has been shown by Fourier to be as constant as are the dimensions
of its orbit, and the period of its annual revolution. Laplace observes,
that our planet has undergone no Contraction of Size during the last
2000 years ; consequently there has been no sensible *Refrigeration*
during that period, and the last Seculum of M. de Beaumont has
already extended to more than twice the length of a Millennium.

Another argument, or rather postulate, has been adduced in favour of *Central Heat,*—the Fusion of Unstratified Rocks, and their
forcible Injection into the Stratified.

Gentlemen, I have confessed to you again and again, that I am not
aware, nor has any one as yet informed me, by what test Stratified
and Unstratified rocks can be distinguished ; the only test I know is
the good will and pleasure of those who make the distinction. The
followers of Pluto seize and appropriate to his use as many rocks as
they think proper. By virtue of such seizure, these Rocks become necessarily Unstratified : why so ? because if Stratified they would be no
longer Plutonic. Stratification I know is a question to be determined
not by the senses but by the fancy ; otherwise, I would say, that the
magnificent range of basaltic cliff, which extends from the county of
Derry along the coast of Antrim as far as Fairhead, is as distinctly
stratified as any mountain-limestone, oolite or chalk in Great Britain.

However, I waive this objection as it leads me away from my subject, and return to the consideration of *Central Heat.* Have those
who believe in this agent ever taken into their account the nature
of the substances said to have been fused ? Many of the trap rocks,
not all of them, (for the family is large, and many of its members have
been introduced into it, not by nature but by adoption,) I attribute
to the agency of the causes which have produced lava, causes which,

* See Comparative View of the Huttonian and Neptunian Systems of
Geology.

comparatively speaking, I do not believe to be very deep-seated. These rocks I put out of consideration for the present; the remark about to be offered apply to granite and its congeners, under which head I would give to every one full liberty to include or reject quartz rock, gneiss, mica slate, eurite, cipollino, hornblende rock, serpentine, &c. Some or all of these, it is the bounden duty of *Central Heat* to fuse and to eject.

Such and so limited are the means of Chemistry, that of many substances thus brought within the sphere of our inquiries, the point of fusion is at this day unascertained. The author of the masterly publication before adverted to, brought together many useful observations upon this subject. He observes that " Lavoisier could not melt a " particle of Carbonate of Lime by the intense heat of a burning mir- " ror, and that Quartz, according to Saussure, requires for its fusion " a temperature = 4043° of Wedgwood's pyrometer, Glass requiring " at a medium only 30° of the same scale."

That the Difficulty, which here suggests itself, of providing, in the absence even of imaginary fuel, a Supply of imprisoned Heat sufficient to fuse the substances I have mentioned and others scarcely less refractory, may be mitigated by extending the time employed in the process, or by the aid of compression and other circumstances, I am ready to admit; but, in the most favourable view of the case, the Heat wanted, (when we consider the thickness and extent of these rocks, comprising entire mountains and mountain chains,) must be prodigious ; and I cannot but admire the singular taste of those geological speculators, who, enjoying the free range of the globe, have deposited their Caloric exactly in that spot in which it can be of least use to them. The inconvenience of this distribution becomes still more apparent when it is recollected that fusion is not all that is necessary ; but that, when fused, these substances must be propelled in a determinate direction and with sufficient force, in many instances, to raise the bed of the sea to the height of an Alpine chain. I will not attempt to point out to you the way in which this is accomplished, but confess at once that I do not understand it.

And yet it appears certain that the surface of our planet has become cooler and cooler, from the period when organic life commenced to the tertiary epoch. If this cannot be explained by the Escape of Heat, there remains only one other mode of explaining it,—a continually diminishing Supply. The latter is the explanation offered by Mr. Lubbock. Sir John Herschel, also, has brought into view causes within the range of physical astronomy which, independently of a Loss of Internal Heat, produce a slow but certain Diminution of Temperature on the surface of our globe *. These auxiliaries, however, are insufficient.

* The Baobab-tree of Senegal is supposed by Adanson to have attained the age of 5150 years, and De Candolle attributes to the *Cupressa disticha* of Mexico a still greater longevity. (Lyell, vol. iii. p. 99.)
If these opinions be correct, it seems improbable that any great change either of level or climate can have taken place at these spots within the last 5000 years.

26

Mr. Lyell has offered another solution of the problem, depending not on celestial but terrestrial causes. The chapter that contains it abounds in valuable information and ingenious reasoning; but when the author tells us that * in every country "*the land has been in some* "*parts raised, in others depressed, by which and other ceaseless changes,* "*the configuration of the earth's surface has been remodelled again and* "*again since it was the habitation of organic beings, and the bed of the* "*ocean lifted up to the height of the highest mountains,*" I cannot but wish that he had stated this as an opinion, not as a fact.

All these theories have one defect in common; they do not meet the whole of the case. We have to explain not only the *Cooling gradual* during the long interval that occurred between the formation of the carboniferous beds and the chalk, but also the *Sudden Chill* which followed, and seems to have continued from that time to this. There is yet another element to be taken into account. The coal-beds of Melville Island contain various plants, natives of the country where they are found, and which, if we may trust analogy, require for their healthy growth or for their growth at all, not only tropical heat †, but a tropical apportionment of the periods of exertion and repose. It is a botanical impossibility that such plants could have flourished in a region in which they must have been stimulated by months of continuous Light, and paralysed by months of uninterrupted Darkness. The distribution of Light, therefore, as well as of Heat, must formerly have been different from what it is at present.

To meet this further difficulty, recourse is had to physical astronomy, which gives us the *Precession of the Equinoxes, and a Shifting Axis of Rotation*: but the periodical changes of astronomers are insufficient to explain the phenomena to which I have just drawn your attention. It has therefore been suggested that a greater change may, in the course of ages, have been produced on the axis of the earth's rotation by some foreign cause, say the *Collision of a Comet*.

Such change is undoubtedly possible, but of possibilities there is no end, and we must circumscribe our researches to render them useful. Sir John Herschel gives us no encouragement, therefore, to proceed with this speculation. Mr. Conybeare also dissuades us from it, but by an argument which to me at least appears inconclusive.

His argument, founded upon the lunar theory, is this,—that the internal strata of the earth are ellipses parallel to its external outline, their centres being coincident, and their axes identical with that of the surface. The present axis of the earth must therefore have been its axis from the beginning. It may have been so, yet I should like to be told by what process the form of the internal strata of the earth had been so nicely determined. Possibly, however, I may not understand the expression "internal *Strata*." All I believe to be ascertained is, that of corresponding sections of the interior the density is

* Principles of Geology, vol. i. p. 113.
† Since this passage was written, doubts have been expressed whether the specimens of these plants preserved at the British Museum are sufficiently distinct to warrant the inference.

nearly the same, and if so, my inference is, not that the earth has never changed its axis of rotation, but that if it has done so, the interior was then sufficiently pliant to accommodate itself to the change. A much more formidable objection to the employment of such a cause is, that if once called in, we must take it with all its consequences. The effects produced by it will not be what we wish performed, but what its nature obliges it to perform. In explaining the phenomena of Melville Island, it might render inexplicable those of the rest of the world. If we choose to change the axis upon which the earth revolves, let us at least fix upon the best time for doing it; now what is that time? immediately after the formation of the carboniferous series? The reduction of temperature at that epoch was inconsiderable; tropical plants and animals are found in the lias, in the oolite series, in the chalk. A much more convenient time would be on the first appearance of the tertiary rocks; but however satisfactory it might be to trace to such a cause the violent changes and disturbances which appear to have taken place about that period in all other parts of the world, I am afraid our satisfaction would be greatly diminished on finding that Gosau and Maestricht * escaped unhurt.

Be the cause what it may, the effect is certain. The Temperature of the Crust of the Earth must have been higher when the Coal-measures were deposited than now, and we have reason to think it was still higher at antecedent periods. That a considerable degree of Heat still exists, either partially or generally, at no great distance from the surface, appears from thermal springs and volcanoes.

I am aware that the doctrine of *Internal Cavities* has been regarded as visionary; and in the extent to which it was carried by some of the old Cosmogenists it was so; but that comparatively near to the surface, there are, I do not say Vacuities, but large Spaces unoccupied by solid matter, is not only probable, but almost proved. It seems, indeed, to be a necessary consequence of the structure of the crust of the earth. No miner has ever got to the bottom of a vein, and a vein itself is often a half empty pipe or fissure. The correspondence of the phases of distant volcanoes, the continuous ranges of their eruptive openings, the vast extent of territory shaken simultaneously by their convulsions, are so many proofs of communication below the surface. The bulk of the ejected matter cannot be less than that of the concreted ejections which we see; for at the temperature of fusion it is greater than at a lower temperature, and for every foot of matter ejected, it is necessary to provide a substitute in the place which it occupied.

The continuous streams of lava which issued in Iceland, on one occasion, attained the length of forty or fifty miles. But the bulk of volcanic matter presented to view, does not enable us to form a correct estimate of the quantity of matter ejected; we must take further into account the combustible substances which have vanished, the gases which have escaped, the dust and ashes which, projected into the air, have fallen many miles distant from the place of explo-

* See the descriptions of these places in Geol. Trans.

sion *. Then only can we entertain a just idea of the Cavities that must
have been created in the interior of the earth by the escape of a mass
of matter competent to produce an Etna or a Chimboraço. Such Cavities are ill suited to support such Mountains ; La Metherie therefore
supposes Cavities to be at a distance, and volcanic matter to flow
from these through long galleries and fissures of communication. Nor
have we in volcanic countries alone decisive evidence of the existence
of subterranean Cavities. No rock is exempt from Fissures : in thick
beds of limestone Fissures and Caverns are exceedingly abundant; and
the extent of these last is sometimes prodigious. Who has not heard
of the Grotto of Antiparos ? of the Caverns of Carinthia and Carniola,
the content of which amounts to some hundred thousand cubic feet?
of the Kingston Cave recently explored near Michelstown in Ireland ?

To the frequency of Caverns and Openings, by whatever name designated, I ascribe many of the inequalities which vary the surface of
the earth ; such openings, I conceive, produce phenomena sometimes
of Subsidence, sometimes of Elevation. I cannot entertain a doubt,
that many of the tilts and contortions of strata usually ascribed to *Sou-
lèvement*, have been occasioned solely by want of adequate support.

The Duchy of Finland exhibits an endless series of lakes filling up
the hollows of a granitic surface. Let me be allowed a similar series
of subterranean lakes occupying similar basins beneath the level of the
Baltic, and receiving, by means of Fissures extending up to the summits of the Scandinavian chain, a continual supply of water which has
no outlet ; in other words, let me be allowed the use of hydrostatic
pressure ; and without having recourse to central heat or secular refrigeration, I think I shall be able to account, without difficulty, not by
a general and uniform Rising, but by a number of unequal and partial
Risings, for the phenomena observed along the shores of the Baltic.

Steam is often referred to as capable of producing the same result,
nor will I deny that it might do so under favourable circumstances ;
but I apprehend Steam rarely does act in nature under such circumstances ; for its existence depends on the access of heat, and its force
on close confinement, contingencies not very likely to occur in the
porous and fissured strata of the earth. Any of the various Gases, if
compressed, might also become agents of elevation, but only under
the same conditions as steam.

I have reserved for the last the popular theory which accounts for
Elevation by the forcible *Inroad of igneous rocks into sedimentary*.

To put this theory to the test, it is natural to inquire, what igneous
rocks are. My answer is, whatever geological speculators think proper
to call so. The late Professor Dugald Stewart cautioned us strongly,
though, alas ! in vain, to avoid the language of theory. Appearances,
he observes, " should always be described in terms which involve no
" opinion as to their causes. These are the objects of separate ex-
" amination, and will be best understood if the facts are given fairly,

* In 1783, a submarine Volcano off the coast of Iceland ejected so much
pumice that the ocean was covered to a distance of 150 miles, and Ships
were considerably impeded in their course.

" without any dependence on what should yet be considered as un-
" known ; this rule is very essential where the facts are in a certain
" degree complicated."
In dealing out to rocks the appellation of *igneous,* some geologists
are more liberal than others. I have not time to enumerate the various rocks which enjoy this title, still less to investigate their respective claims to retain it. I will therefore content myself with observing,
that in the scantiest catalogue they are many in number, and consequently, if ejected in a state of fusion, must have been ejected from
different reservoirs and cauldrons, not from a *central* cauldron.
That any rock whatever was originally igneous, is a gratuitous assumption. Lavas themselves may be, and probably are, in very many
cases, Rocks not originally igneous, but Rocks which have been exposed at one time or other to the action of fire.
Granite is one of the rocks most usually considered as an *Agent in
Elevation,* for what reason I am at a loss to discover. Solid Granite
has no inherent principle of motion ; if it move, it can only be by virtue of the impulsion it has received from some other body, not in consequence of its igneous origin or its want of stratification. The disturbances of strata that adjoin granite are not more constant, nor more
striking nor more extensive than those of strata far remote from it, as
for instance, the limestone shales of Derbyshire or the coal-beds of
Liege. Granite veins are too small to raise mountains, and the changes
or anomalies that take place at the junction of granite with other
rocks, whatever else they may prove, appear to me to have no bearing on the question of *Elevation.* On the other hand, the arguments
adduced against the doctrine that Granite while fluid has been forcibly
injected from beneath into its present position, are to my mind conclusive; especially that which is founded on the frequent transition
which takes place from Granite to the rocks that adjoin it. We find a
continuous series from Granite through Gneiss and Mica slate to Clay-
slates and the Fossiliferous Slates ; and it is not possible to stop at any
point of this progress, and to say in which direction the tendency is
strongest. If the gradation were single, the difficulty would be great,
but what shall we say to a repetition of such gradations ? In Mr.
Weaver's paper on the East of Ireland, two detailed sections are
given, in one of which, more than six layers of Granite alternate with
as many of Mica slate, and in the other five alternations of the same
kind occur, the rocks in each instance forming bands from three to
seventy fathoms in thickness.
The reliance which some authors place on Granite and other unstratified rocks, as *Agents of Elevation,* is to me very extraordinary;
let one instance suffice. At Castrogiovanni in Sicily, the Pleiocene
Beds attain an altitude of three thousand feet ; hence it has been inferred, that *since these beds were deposited, there has been formed and
introduced into the beds subjacent, a body of Granite, Sienite, Porphyry
or other crystalline and unstratified Rocks three thousand feet in thickness.* This supposition is said to be necessary, but since I do not see
the necessity, I will venture another supposition, viz. that Etna has
not risen to the height of ten thousand feet without occasioning large

cavities in its neighbourhood, some of them submarine; that Castro-giovanni is situate over one of these; that the Pleiocene strata have closed the cavity and rendered it water-tight, except on the side of Etna; from whose lofty flanks and cloud-capped crater the caverns beneath are regularly supplied by fissures with rain-water and melted snow. Let the author grant me so much,—I ask no more, The hydrostatic paradox has tripped up the hills of the geological one, and I behold my Pleiocene beds mounted at once on a pedestal three thousand feet high, and capable of still further promotion.

If the explanation here offered meets the case of Castrogiovanni, it will equally account for the height of the tertiary beds in different parts of the Val di Noto, and for similar phenomena in every country which is or has been formerly the site of volcanic eruptions.

To the appearances on the Gulf of St. Lawrence, described by Captain Bayfield, I have already adverted.

My Predecessor directed your attention last year to the existence in the Morea of four or five distinct Ranges of ancient Sea-cliffs, marked at different levels in the limestone escarpments by lithodo-mous perforations, lines of littoral and sea-worn caverns, and other striking proofs of former tidal action. Similar Terraces have been ob-served in Sicily, in Chili, in the Gulf of St. Lawrence and various other places. At Uddevalla in Sweden, are ancient Beaches with shells of living species, two hundred feet above the level of the Baltic, a height strikingly disproportionate to the very moderate Rise ascertained to have taken place in other parts of the Scandinavian coast: many ex-amples of similar phenomena have been found in Great Britain. It would be rash to offer a solution of these phenomena in the gross. Every individual case deserves separate examination. All I undertake at present is to put a new key into the hands of the decipherer.

It was my intention on commencing this address to have discussed at some length the theory of M. Elie de Beaumont, but there is not time now to do it justice. He belongs to that class of authors whose opinions, right or wrong, always instruct me. There is no part of his theory which does not evince thought and diligence, a habit of cor-rect observation and an enlarged mind. In some respects I differ from him, and it will not be difficult to infer from what I have already said, wherein the difference consists. Should these observations engage his notice, I would beg him to consider whether the distur-bances in the Alps and elsewhere have not been generalized rather more than they will bear, whether the tilts and upliftings may not have taken place bit by bit at various epochs, and whether, if the *se-cular Refrigeration of the Globe* cannot be established, and *Central Heat* be an *Ignis fatuus*, his attention may not be usefully directed to more partial but better authenticated sources of disturbance and elevation.

Allow me, in conclusion, to say a few words upon a subject in con-nexion with which my name has of late been brought forward much more prominently than I could have desired;—I mean *Diluvial Action*. Some fourteen years ago I advanced an opinion, founded alto-gether upon physical and geological considerations, that the entire earth had, at an unknown period, (as far as that word implies any

determinate portion of time,) been covered by one general but temporary Deluge. The opinion was not hastily formed. My reasoning
rested on the facts which had then come before me. My acquaintance
with physical and geological nature is now extended; and that more
extended acquaintance would be entirely wasted upon me, if the
opinions which it will no longer allow me to retain, it did not also
induce me to rectify. New data have flowed in, and with the frank-
ness of one of my predecessors, I also now read my recantation.

The varied and accurate researches which have been instituted of
late years throughout and far beyond the limits of Europe, all tend to
this conclusion, that the geological schools of Paris, Freyberg and
London have been accustomed to rate too low the various forces
which are still modifying, and always have modified, the external
form of the earth. What the value of those forces may be in each
case, or what their relative value, will continue for many years a
subject of discussion; but that their aggregate effect greatly sur-
passes all our early estimates, is I believe incontestably established.
To Mr. Lyell is eminently due the merit of having awakened us to a
sense of our error in this respect. The vast mass of evidence
which he has brought together, in illustration of what may be called
Diurnal Geology, convinces me that if, five thousand years ago, a
Deluge did sweep over the entire globe, its traces can no longer be
distinguished from more modern and local disturbances. The first
sight of those comparatively recent assemblages of strata, which he
designates the *Eocene, Meiocene* and *Pleiocene* Formations, (unknown
but a few years ago, though diffused as extensively as many which
were then honoured with the title of universal,) shows the extreme
difficulty of distinguishing their detritus from what we have been
accustomed to esteem Diluvium. The Fossil Contents of these formations strongly confirm this argument. M. Deshayes has shown
that they belong to a series unbroken by any great intervals, and
that, if they be divided from the secondary strata, the chasm can
have no relation to any such event as is called The Flood.

Further, the elephants and other animals once supposed to be exclusively *Diluvial*, are now admitted to be referrible to two or three
distinct epochs; and it is highly probable that the blocks of the Jura
Mountains, of the North of Germany, of the North of Italy, of Cumberland, Westmorland, &c., are not the waifs and strays of one, but
of several successive Inundations.

It is, Gentlemen, a well-known rule of such institutions as ours,
that the "Authors alone are responsible for the facts and opinions
contained in their respective productions." Under that feeling have
I spoken on the present occasion, and having freely set before you
what has occurred to me on some points of general interest to our
science at this time, I think it my duty, in concluding this address,
to disclaim and deprecate any attempt to connect what I have here
expressed with the general sentiments of the Geological Society. The
opinions I have uttered are my own, and I should be sorry that
more importance should be attached to them than they intrinsically

deserve, from the accident of their having been delivered from this Chair. Had not the whole responsibility fallen on myself, I should have hesitated, or perhaps altogether forborne to bring before you Opinions, several of which I know are little in accordance with those of some of the most distinguished members of our association.

Printed by RICHARD TAYLOR, Red Lion Court, Fleet Street.

ADDRESS

DELIVERED AT

THE ANNIVERSARY MEETING

OF THE

GEOLOGICAL SOCIETY OF LONDON,

On the 20th of FEBRUARY, 1835;

By GEORGE BELLAS GREENOUGH, Esq., F.R.S.

PRESIDENT OF THE SOCIETY.

———

LONDON:

PRINTED BY RICHARD TAYLOR,

RED LION COURT, FLEET STREET.

———

1835.

ADDRESS, &c.

GENTLEMEN,

I REJOICE to see you assemble in these long-desired apartments: the increased facilities now afforded to us of pursuing the objects of our institution demand on our part a corresponding increase of exertion. Let it not be imagined that in accepting a boon from the Government we have not incurred an obligation. Our claim rested entirely on the sense which the public entertain of our deserts. The full and accurate knowledge, which it has been our aim to acquire and publish, of the subterranean resources of all parts of the world, cannot fail to be useful: so long as the utility of our establishment is distinctly felt, independently of its bearings upon what to us appear the higher considerations of science, so long may we rely on the continuance of public support; but our claim to that support would vanish at once, if we should relax our exertions, and fail to realize those expectations which our hitherto well-sustained activity has kept alive in the breast of our benefactors.

Gentlemen, you are well aware that all the accommodation which you now enjoy has not been derived, however, solely from the beneficence of the Government, set in motion by the Royal Society. On our first taking possession of this house the repairs and alterations occasioned on our part an immediate outlay of 1500*l*., and 1000*l*. more have now been expended in improving and furnishing the new apartments, and in availing ourselves of such other advantages as their acquisition has enabled us to command. The Council considering it imprudent that so large a sum should be deducted from the capital of the Society, 500*l*. of that amount have been provided by a voluntary subscription.

The alterations planned by Mr. Decimus Burton have been executed under his direction within the specified time, and in accordance with the estimates. With what skill they have been executed has not escaped the notice of the Building Committee; with what success is apparent to you all. You have been informed that Mr. Burton has declined to accept any remuneration for his professional services, but I cannot deny myself the pleasure of again recording this new instance of his public spirit and characteristic liberality.

The concluding part of the third volume of our Transactions is in the press, and the recent arrangements of the Council induce me to

A

hope that not this part only, but also the first part of the fourth volume will be ready for delivery in the course of the present year.

The Number of our Fellows has received an addition of forty-two
and a diminution of ten ; of those whom we have lost, three only are
known to me as Contributors to the Transactions or as Geological
Authors.

To the liberality of Mr. Matthew Culley, the chief of a family
greatly celebrated for the practical improvements they have introduced into agriculture, our Museum is indebted for a large series
of primitive rocks collected in the remoter parts of Sutherlandshire, as also for some fine specimens of the fossil fishes which occur
at Banniskirk, situated eight miles south of Thurso in Caithness.
The geological relations of these last have been since investigated
by Mr. Murchison and Professor Sedgwick, and their zoological
characters determined by M. Agassiz*. Mr. Culley also transmitted
to us an account of the prodigious power occasionally exerted by
rivulets when swoln by heavy rain.

Major James Franklin was a younger brother of Captain Sir John
Franklin, R.N.; he commenced the survey of Bundelcund in India
in 1813, and continued it during four years, after which he joined
the army in the field, under the Marquis of Hastings. In December
1818, Captain Franklin returned to the duties of his survey, and
shortly afterwards was promoted to the rank of Assistant Quarter-
Master General. In 1820, he commenced at Calcutta the construction of his maps. He afterwards made a careful survey of Sincapore
and the adjacent strait. Having repaired to England in 1823 for
the benefit of his health, he soon learned by communication with
members of this Society, how much the value of his surveys would
have been enhanced by geological descriptions. On his return to
India in 1826, his first care was to supply this desideratum : he
solicited, though unsuccessfully, the appointment of Geologist to
the trigonometrical survey then carrying on in India, and strenuously recommended that all officers employed on that service should
be qualified and encouraged to collect materials for the construction of a geological map of the entire Peninsula. Declining health
brought him again to England in 1829, where he remained till
his death, which took place in the summer of last year. He transmitted to us a paper on the geology of a portion of Bundelcund †
and other districts in Central India; and to the Asiatic Society of
Bengal a description of the Diamond Mines of Panna, published in
the Transactions of the respective Societies. Among his MSS.
have been found Observations on several Iron Mines, and on the
mode in which the ore extracted from them is manufactured in
Central India, together with an account of different beds of coal
in that country.

Mr. James Hardie evinced in early life a taste for natural history.
Educated at Edinburgh, he founded the Plinian Society in that

* Proceedings of the Society, May 1829.
† Transactions of the Geological Society, 2nd Series, vol. iii. Part I.

city, and contributed largely to its Museum. In 1784 he embarked for India, and served in the Bheal campaign; he was afterwards appointed the Residency Surgeon at Odeypoor, and made a survey of the neighbourhood. In his visits to Calcutta he acquired the friendship of the most distinguished geologists of the East; he became a member of the Medical and Physical Society, and contributed many papers to their Transactions. A survey of his route on one occasion from Calcutta to Bombay, and thence to Odeypoor, will be found in the Transactions of the Asiatic Society, of which he was also a member. In 1830, he made a voyage round the Indian Archipelago, with a view to the recovery of his health. He passed six months at Java, and paid much attention to the geology of that island. On his return to Scotland he presented his collection to the Museum at Edinburgh. Professor Jameson recommended him to the East India Company as a fit successor to the appointment held by the late Dr. Turnbull Christie. Mr. Hardy repaired to Paris in 1833 for the purpose of prosecuting his geological studies, and died there in May following at the age of 31.

In reviewing the geological labours of the year I shall advert principally, but not exclusively, to those of our own members, and the order of precedence will be regulated by the nature of the respective papers, without any reference to date.

Mr. Murchison, in prosecution of the work in which he has been so long and actively engaged, has communicated to us his observations on the detritus that covers the old red sandstone in Herefordshire and its vicinity. All the detritus, he says, seems to be derived from neighbouring rocks. Granite boulders are nowhere found within its area, but they occur of large dimensions and of various sorts upon its northern confines; he states generally that they appear to have come from the North. Many, if not all of them, may I believe be identified with the granitic rocks of Westmoreland and Cumberland. Several of those I have observed on the north of Shrewsbury have the character of the Irton rather than of the Shap granite. The detritus of the old red sandstone is ascribed to the operation of different causes, some of which may perhaps require further study.

From Mr. Strickland we have received three communications respecting certain bones of extinct quadrupeds associated at Cropthorne in Worcestershire, with existing species of shells. On a base of lias clay reposes a layer of fine sand containing twenty-three species of land and freshwater shells, together with rolled and broken bones of the Ox, Deer, Dog, Bear, and Hippopotamus. Upwards this sand passes into gravel undistinguishable from the so-called diluvium. These shells are found at five or six different localities within the Vale of Evesham. Two of the species are thought to be extinct. The inference drawn from all the phænomena is that this deposit occupies the site of a former river-bed or lake; that since its formation mammiferous animals have migrated more than molluscous, and that the climate has remained nearly stationary.

Mr. Edward Charlesworth has placed in our Museum some valu-

able specimens collected at Sutton in Suffolk, valuable because they
establish the existence of similar phænomena there also. The specimens consist of bones, (one of which appears to belong to the elephant,) and of freshwater shells, including an extinct species of Cyclas.
Five or six feet below the surface a layer of calcareous nodules may
be traced, he says, for about half a mile along the banks of the
Stour ; the nodules are very numerous and found only with the
shells. Let us hope that the interest excited by these notices may
lead some competent naturalist to undertake a detailed examination
of the " Crag," our imperfect acquaintance with which, again and
again pointed out from this Chair, continues to be what I may justly
call one of the Opprobria of English Geology.

The changes which have taken place in the boundaries of the
land and sea on the north-eastern coast of Kent have been brought
under our notice by Mr. Richardson.

Mr. Wetherell has transmitted to us a detailed account of a Well
recently sunk on the south side of Hampstead ; its depth is 330 feet;
under 285 feet of clay was found a rock five feet thick, which in external character and fossil contents agrees with the Bognor, and
rests on plastic clay. At the depth of 160 feet specimens were
brought up of what would be called in Sheppy Island fossil fruits :
the local distribution of these remarkable bodies is very limited.
From the examination of Mr. König, it would appear that a few only
are of vegetable origin. By far the greater part belong to a distinct
class, which, in Mr. König's new division of the natural order of
Polypi, are distinguished by the appropriate name of Carpomorphi.

Mr. Rofe has found in the neighbourhood of Reading the Bognor
rock, occupying the same position which Mr. Wetherell assigns to
it at Hampstead. It is strange that in a quarry of which so many
sections have been given, this rock should have remained so long
unnoticed. Mr. Rofe has also observed, that in the wells at Reading
the level of the water depends not on the Kennet which is the nearest
river and flows over tenacious clay, but on the more distant Thames
which flows over gravel resting upon chalk.

M. Boué has recently pronounced the plastic clay formation
destitute of fossils, but it certainly does contain them in England.

Mr. Woodbine Parish has detected chalk in a part of the Sussex
coast, where its existence had not been before observed. He has
traced it from Felpham to the distance of about a mile ; it runs in
the direction of Middleton, where chalk marl has been obtained at
low water.

Dr. Mitchell has pointed out to us in what respect the chalk of
the North of England differs from that of the South. The difference
consists in its greater hardness, its occasional redness, its well defined stratification, the absence of flint nodules in its upper portion,
and the continuity of layers of flint in its lower; in many of these
characters it resembles the chalk of Antrim.

The occurrence of Hippurites in the chalk of Sussex appears to
require confirmation.

A well recently sunk at Diss in Norfolk, after penetrating two beds of clay and sand, the aggregate thickness of which amounted to 100 feet, penetrated through the great body of the chalk, which proved to be 500 feet in thickness; the tools passed afterwards through five feet of sand, when water flowed in, and rose to within 47 feet of the surface. Mr. Taylor, our Treasurer, to whom we are indebted for this information, states that the proprietor was encouraged to persevere in so expensive an undertaking by the account which he had seen of a similar well in Cambridgeshire, drawn up by Mr. Lunn and published in our Transactions. The confidence which he placed in that analogy (though the places are many miles distant from one another,) has been fully justified by the event, and the success of the experiment furnishes a striking example of the value of records of this nature.

Dr. Fitton's paper on the strata of the South-east of England, between the chalk and Oxford oolite, together with the memoirs of Dr. Buckland and Mr. De la Beche on the coast of Dorsetshire, are in great part printed.

The Wealden and Purbeck beds are generally supposed to have been deposited in an estuary, but what may have been the form, direction and extent of that estuary, it is not easy to conjecture. Similar beds occur at Lady Down in Wiltshire, and at Swindon; and the Vivipara, one of their characteristic fossils, appears towards the top of Shotover. The discovery of this formation in the neighbourhood of Beauvais excited much interest among the naturalists, assembled there about three years ago, and several species of the Wealden fossils seem to prove its existence at Loch Staffin in the Isle of Sky.

And here, while considering the district of the Weald, allow me to direct your attention for a moment to a subject temporarily, and let us hope, only accidentally connected with it. I refer to the repeated, though slight shocks of earthquakes, which have been experienced during the last eighteen months in the neighbourhood of Chichester. When we recollect the prodigious area over which the earthquake of Lisbon was felt, comprehending one fourth of the entire northern hemisphere, we can hardly suppose it possible that the cause of such phænomena can be seated so near the surface of the earth as to come at all within the range of our observation. It is, however, a singular coincidence, that in 1734, 1747, 1749, 1755, and the three following years, earthquakes were felt in this same part of England, and that on those occasions, as well as now, the direction of the shocks is supposed to have nearly coincided with the great line of fault.

On the oolite district we have received no additional intelligence. The boundaries of the several clay-beds in the midland and northern counties is as yet ill defined, and the same remark will apply to many of the stone-beds in Lincolnshire, Leicestershire, and Rutlandshire. The anomalies of the Stonesfield slate are still unexplained, and it is very desirable that figures should be published of the fossil remains found in it, and more especially the coleopterous insects.

Mr. Murchison has discovered an outline of lias at Longdon Hill, near Upton, and improved the boundary line of that formation in the neighbourhood of the Severn. The lias marl or lower lias rock, ill exhibited along the Yorkshire coast, and on the coast of Dorsetshire imperfectly developed, in Worcestershire and Gloucestershire swells into importance, and may be traced continuously for a distance of twenty or thirty miles, passing insensibly here, as in Germany, into the red marl.

A much more unexpected discovery has been made, by the same indefatigable observer, of the same beds in another situation, viz. between the Hawkstone Hills and the towns of Whitchurch and Market Drayton, seventy miles apart from the lias range in the Midland counties. This outline is known to extend about ten miles in length, and from four to six in breadth; its greatest extent is from north-east to south-west; its western boundary is obscure. The strata have a slight dip inwards towards a common centre. The visit of a geologist to this spot has had the effect of stopping an inconsiderate speculation; the bituminous character of the lower beds had been supposed to indicate the proximity of coal, and a shaft had been sunk to the depth of three hundred feet in search of that mineral, in a place which I need hardly say afforded not the slightest chance of success.

The escarpment here exposed presents to view the junction of the middle and lower part of the formation. With most of its fossils we are familiar, but six or seven of the species met with are new in England, though some of these occur in a corresponding position at Brora in Sutherlandshire, and others are figured in the valuable work which Mr. Zieten is now publishing on the Organic Remains of the Kingdom of Wurtemburg. A shaft sunk at Kentsrough has reached the brine springs and gypsum of the subjacent formation.

Mr. Williamson, jun., has given us a detailed account of the lias near Scarborough, to which I shall have occasion to advert again.

Mr. Murchison has described to us the new red sandstone on the confines of England and Wales; the formation here, as in the north-eastern counties, may be divided into

1. Red and green marl;
2. Sandstone and conglomerate;
3. Calcareous and dolomitic conglomerate;
4. Lower red sandstone:

corresponding to 1. the Keuper; 2. Bunt sandstein; 3. Zechstein; 4. Rothe todte liegende of the Germans.

Limestone is sparingly distributed in this district. No trace has been met with of the muschelkalk, and the magnesian limestone occurs only in the shape of a sandstone conglomerate.

1. The red and green marls afford brine-springs in Gloucestershire, Worcestershire, Shropshire, and Cheshire, but with a small proportion of gypsum.

2. The beds immediately beneath are largely developed in the north of Shrewsbury, in Staffordshire and Salop; the district which they occupy is wild and barren, owing to the prevalence of decom-

posed quartzose conglomerates. Where sand predominates it is more fertile, and the country assumes a character perfectly distinct from that which belongs to the upper and lower beds.

3. The fragments of the calcareous conglomerate are occasionally of oolitic limestone, sometimes of old red sandstone, or of some member of the coal series. This division is unimportant on the east of Coalbrook Dale, but in the western part of Salop it swells out and corresponds with the dolomitic conglomerate of Bristol. At Cardeston it is from eighty to one hundred feet in thickness ; its escarpment, with partial interruptions, may be traced from Alberbury and the Brythins round the carboniferous promontory of Salop and the Clent Hills, and it forms a distinct ridge between Bridgenorth and Kidderminster.

4. Beneath this, in Salop and Worcestershire, is found a thick deposit of reddish sandstone (rothe todte liegende); it passes upwards into dolomitic conglomerate, and downwards into the coal-measures in conformable beds, so that there is great reason to suppose that unwrought coal lies beneath. Mr. Murchison has determined the extent of this rock, which occasionally towards the bottom contains trappean conglomerates like those of Heavitree in Devon, or feldspatic rocks like those of the Malvern and Abberley range.

The dolomitic conglomerate just mentioned has in another county engaged the attention of the Rev. David Williams : he has discovered for the first time Saurian reptiles in this deposit. Mr. Conybeare had before noticed the occurrence of a part of the skeleton of a supposed gavial near the bottom of the red sandstone in Worcestershire.

The obscurity which for so many years continued to involve the red sandstone deposits both in England and on the Continent was first cleared away by Professor Sedgwick. We now know that the proper position of the rothe todte liegende of the Germans is immediately beneath the magnesian limestone, and that it is the same rock which Mr. Smith described as the Pontefract sandstone. It is much to be regretted that Mr. Hoffmann in the beautiful map and sections which he published of North-western Germany, has designated by one name, as if they were only parts of one and the same deposit, the red sandstones and conglomerate that lie under the carboniferous series, and those that lie above it. This grouping together of formations so widely separated in nature is very objectionable. With a view to distinctness it is essential that the rothe todte liegende shall not be classed as heretofore, sometimes with the old red sandstone, sometimes with the new, sometimes with both, sometimes with the coal-measures, but that it should hold the rank of a substantive and independent formation. The barbarous phrase, which has just been employed to designate it, though tolerated in Germany, will never, it is hoped, be naturalized here. The name by which it is known at Tarnowitz is too local. Many other names have been proposed when one would have sufficed. By M. de Beaumont it is called Grès des Vosges; by Mr. Smith, the Pontefract rock ; by Professor Sedgwick, the Plumpton ; and by Dr. Hibbert, the Roslyn sandstone. Assuming that all these names are synonymous, the last appears for many and obvious reasons the most worthy of adoption.

This important rock, which, in richness of soil, in undulation of surface, and in the luxuriant growth of its timber, closely resembles in many places the old red sandstone of Herefordshire, lies in general, sometimes conformably, sometimes unconformably, upon coal-measures, and even contains occasionally beds of that substance. But in the neighbourhood of Shrewsbury, and also at Tasley and Coughley near Bridgnorth, Mr. Murchison has shown that this lower red sandstone overlies unconformably and passes down into a zone of coal-measures containing a peculiar fresh water limestone. The great coal-beds of Brozeley and Coalbrook Dale are wrought beneath it.

I had occasion myself to observe during the last summer, in the neighbourhood of Nuneaton in Warwickshire, a limestone similar in aspect, lying under a two-foot bed of sulphureous coal. The limestone is about fourteen inches in thickness, and exhibits veins of galena in calcareous spar. At the first pit I visited there was but one bed of limestone, but at another on the same estate is a second bed which also contains galena, and on its surface numerous impressions of plants. The interval between the two beds is occupied by a sandstone not unlike the Pennant rock in appearance, and what is here called a chance coal. Immediately beneath the lower one is another bed of coal four feet in thickness.

In three papers which Mr. Conybeare has lately published on the relations of our principal coal-fields*, he considers it probable that coal will be discovered hereafter in many districts as yet unexplored. He dwells upon our uncertainty as to the boundary of the carboniferous beds in the midland counties, and recommends that a survey should be undertaken expressly with a view to determine this problem : " It is little to the credit," he observes, "of a nation like ours, so peculiarly dependent on this branch of her mineral resources, that we continue to acquiesce in a state of ignorance so easily removed. We here see a strong instance of our want of a regular school of mining, such as is possessed by many countries."

Mr. Elias Hall has published a geological map of Lancashire, a county hitherto comparatively neglected, and, I am sorry to add, very indifferently represented in all the geological maps of England. Mr. Hall is entitled to great praise for his intrepidity and perseverance ; had he not possessed these qualities in an eminent degree, he never would have entered, as it were alone and single-handed, on so irksome and laborious an investigation. That the work is in many respects imperfect must be admitted, but considering the apparent disproportion of his means to his end, it is surprising that the author should have achieved so much : what he has left incomplete

* Lond. and Edinb. Phil. Mag. and Journ. of Science, vol. iv. pp. 161 and 346, vol. v. p. 11.

or inaccurate will be readily supplied and corrected by the supplemental labours of more fortunate observers, when the physical features of this extensive tract shall have been accurately delineated by the Ordnance Department. For a detailed account of the carboniferous tracts in Salop and the adjacent counties we are indebted to Mr. Murchison. The following are the conclusions which his paper tends to establish.

1. In the Shrewsbury coal-field the presence of a younger series of coal-measure than has hitherto been noticed, characterized by the freshwater limestone above alluded to.

2. The recurrence of these beds at Coalbrook Dale, over an older series of coal-measures which at one spot repose on mountain limestone, at other places either on the old red sandstone or on transition rocks.

3. The absence of these upper beds at the Titterston Clee hills, where the lower beds rest in two places on mountain limestone, but generally on old red sandstone, as they do invariably on the brown Clee hill, in the forest of Wyre and at Newent.

4. In some of the poor and ill-consolidated coal-beds, particularly in the upper part of the series, the characters of the fossil plants, both generic and specific, can be recognised in the coal itself.

5. The mountain limestone where it does occur in this part of the country is of inconsiderable thickness, and wedge-shaped, so that it shortly disappears entirely. Its absence, therefore, is not to be imputed to mighty convulsions, but to partial and scanty deposition in the first instance.

At Shaftoe, near Wallington, in Northumberland, Mr. Trevelyan * has observed among the constituents of the millstone grit, the lowest bed of the regular coal-measures, transparent fragments of garnet ; they occur there rather abundantly. He has also remarked in other northern coal-fields small portions of hornblende in a similar situation.

The Rev. Mr. Williamson has directed attention to certain ravines in the Mendip hills, and other heights which bound the coalfield of Bris ol. These ravines cross the ridges transversely so as to connect the opposite valleys, being occupied in part by horizontal beds of dolomitic conglomerate and lias ; he infers that the fractures took place before these rocks were deposited, and that the bone-caves were formed at the same period.

Dr. Lloyd first observed fossil fishes in the old red sandstone. Mr. Murchison finds the observation true over a considerable extent of country ; they belong chiefly to the genus Cephalaspis of Agassiz ; they have also been described by Dr. Fleming, as having been met with in the old red sandstone of Forfarshire. They appear not to be diffused through the formation generally, but to be confined to the middle portion of it, the cornstone.

The nature of the pebbles imbedded in the Old Red Conglomerate varies according to its locality. In Scotland they are very fre-

* Lond. and Edinb. Phil. Mag. and Journ. of Science, vol. vi. p. 76, (1835.)

12

quently of gneiss, as is the case in the neighbourhood of Baden-Baden.

In an Account of the Trap Rocks of the Border Counties, and their effects on the Stratified Beds with which they are in contact, Mr. Murchison has separated the objects of his examination into three classes: 1. the Trap-Rocks which penetrate transition beds ; 2. those which penetrate the old red sandstone ; 3. those which penetrate the coal-measures. He refers the whole to igneous action, and considers them to be of the same age as the rocks with which they are respectively associated, rocks which he readily admits to be sedimentary, since, though composed of volcanic materials, they contain organic remains. The igneous action he conceives has taken place under water, and the finer volcanic ejections, arranged by Neptunian agency, have led to the formation of volcanic sandstones. His views upon this subject appear to be in exact accordance with those of Mr. de la Beche.

Treating of the relations between igneous and fossiliferous rocks, Mr. de la Beche observes*, that though frequently posterior, the former are in many cases contemporaneous with the strata in which they at present occur, appearing to have covered an inferior bed, and to have been subsequently covered themselves by a tranquil deposit of transported matter, as lava may flow over a sandy bottom and afterwards be covered up by sand or mud. Trappean rocks, he continues, are in various parts of Europe much associated with the lower parts of the grauwacke series, sometimes in a manner which leaves no doubt that some of them have not been included among the strata after their consolidation, while others have clearly forced a passage through the grauwacke and previously formed masses of trap. Beds of greenstone or porphyry, he says, sometimes fine off among the grauwacke strata, taking the character of an arenaceous deposit, as if such portions constituted a deposit of trappean ashes, thrown out at the same time that the trappean rock itself was produced. Brent Tor, north of Tavistock, remarkably exemplifies some of these appearances.

The researches of your Vice-President in the counties of Devon and Somerset have been carried on this year with increased energy. Of the eight sheets of the Ordnance Map upon which he has been engaged, four were published last spring, three others are complete, the eighth is nearly complete, and an explanatory memoir with sheets of sections applying to the whole are to be published before our next anniversary. Let us hope that this work so admirably begun may not be suffered to terminate here.

Gentlemen, we had many of us an opportunity of witnessing at the late Meeting of the British Association the increasing interest and success with which geology is pursued in Scotland, and we felt more especially grateful on that occasion to Lord Greenock and the Highland Society, for the exertions which they have re-

* Researches in Theoretical Geology, p. 384.

cently made to unravel the structure of their native land, and more especially the nature of its coal-fields. It is not my intention to detail to you all the proceedings of that Society, but I must not refrain from attributing mainly, if not solely, to their exertions the provision which the Government have lately made for the immediate publication of Dr. MacCulloch's geological map of Scotland. Whatever may be the intrinsic excellence of that work, it must be eminently useful, if considered only as a nucleus, round which will immediately congregate those ample stores of geological knowledge which at present lie latent in the minds and cabinets of our northern brethren. Nor will Ireland be backward in furnishing her contingent. The coloured copy of Arrowsmith's map of that portion of the United Kingdom which Mr. Griffith has undertaken to lay before the British Association in August next, will bring within our reach an abundant supply of geological information, which though it has been in his possession for many years past, a natural repugnance to combining geological correctness with geographical inaccuracy has hitherto induced him to withhold.

The exertions of the Geological Society of Dublin have been continued, and cannot fail to diffuse over the whole of Ireland a taste for those studies which at a very early period reflected so much lustre on the name of Kirwan.

It will be in your recollection that Mr. Weaver presented to us some time since a valuable Memoir on the Geology of the southwestern part of that country. In one part of the Memoir the coalmeasures of the county of Kerry were referred to the transition series ; the correctness of this statement was questioned at the time, and various inquiries were instituted and persevered in, without leading however to any very decisive result. Since the commencement of the session, the author on reexamining the district, has with great candour acknowledged himself to have been in error. More diligent investigation brought into view a well-characterized band of old red sandstone, intervening in one part of the coal-field, between the carboniferous and the transition strata.

Mr. Jephson has transmitted to us an account of a remarkable spring at Mallow in the county of Cork, the temperature of which varies from 67° to $71\frac{3}{10}°$. It breaks out in limestone.

An ample and able account of the recent progress of our science on the Continent will be found in the Report of M. Boué to the Geological Society of France. I shall, therefore, confine my observations on this head almost exclusively to the Papers which have been read at our evening meetings.

The first in order relates to the loamy deposit, called in the valley of the Rhine, Loes, a term as yet scarcely naturalized among us, and which, I believe, is correctly represented by the word Silt. This paper, from the pen of Mr. Lyell, has since been published entire in Jameson's Journal.

Intimately connected with this is a communication by Mr. Horner, on the nature and quality of the solid matter actually suspended in the water of the Rhine. To ascertain them the author made experiments during the months of August and November, bringing up about a gallon of water from different depths and drying slowly the solid matter obtained from it. With whatever attention to accuracy such experiments are conducted, they must, I conceive, be multiplied almost indefinitively before we can arrive in safety at any general conclusion upon so intricate a problem.

From Colonel Silvertop we have received a description of certain tertiary deposits, which in the kingdom of Murcia, in Spain, occupy extensive plains, bounded by discontinuous ridges of nummulitic limestone, transition rocks and mica-slate : the author divides these deposits into four districts, and each of these is separately treated. M. Deshayes refers their imbedded fossils to the second and third deposits of tertiary formation.

In a work on Spain, published during the past year by Captain Cooke, will be found a brief account of the mines and rocks of that hitherto partially examined country.

I may also be permitted to notice among the additions which have been made to our library, an excellent Memoir by M. le Chevalier Albert de la Marmora, on the constitution of the Balearic Islands.

No communication has been made to us from Asia since the last Anniversary.

A paper by Mr. Cunningham describes the physical structure, and to a certain extent, the geological composition of the country between Hunter's River and Moreton Bay, in Australia, and is accompanied by a valuable map and section and a small collection of rock specimens. The additions made during the expedition referred to by Mr. Cunningham are important, and the geological notices, though slight, will be welcomed by future inquirers.

Mr. Rogers, who laid before the British Association at Edinburgh an able sketch of the "Geology of North America," has more recently favoured this Society with an account of the strata situate on the banks of the Missouri and Mississippi rivers, and further, in the district of the Rocky Mountains. It may be said of all these papers that they are in a great degree compilations, but compilations so executed are perhaps among the most valuable documents that can be transmitted to us. No general views could ever be opened if every author were to confine his descriptions and reasonings to those minute tracts which have fallen within the sphere of his own personal examination. Every system and theory is necessarily founded upon details industriously collected from various quarters.

Besides these communications, we have received from America recently two works, in which the same subject is treated with great clearness and in considerable detail: the one entitled "Contributions to Geology, by Isaac Lea, accompanied by six plates of Shells," of which some at least are not very accurately figured ; the other "A Synopsis of the Organic Remains of the Cretaceous Rock, with nineteen lithographed plates of Shells, by Dr. Morton." These works, together with the papers of Mr. Conrad published previously in the American Journal of Science and the Journal of the Academy of Natural Sciences of Philadelphia, illustrated also with lithographic plates, have rendered the upper formations of the United States as intelligible as those of our own country.

Dr. Morton notes the generic accordance of the Testaceous Mollusca on the east and west shores of the Atlantic; but independently of genera, there are at least twenty-four species common to both. In like manner some identities have been traced in the tertiary deposits of Europe and America. The *Pecten quinquecostatus* in particular occurs equally in the cretaceous group on both sides the Atlantic; nor is the analogy confined to Testacea; it extends to the Saurian reptiles. The animals whose remains are found in chalk formerly inhabited the seas of the two continents, and whatever cause bared the eastern, appears to have acted simultaneously on the western mass; not a rush of currents, but a subsidence or elevation.

In the county of Onondago, in New York, is a lacustrine deposit still forming, in which thousands of tons might be obtained of bleached shells. The shells at the mouth of the Potomac river, belonging to the newer Pliocene beds, retain their colours; twentynine of the species are the same with those which now live, and of these there are seven only which are not known to inhabit the coast of America.

From the upper marine deposit of Dr. Morton, which corresponds to the lower tertiary of Mr. T. A. Conrad, and to the older pliocene of Mr. Lyell, numerous specimens were exhibited to us in the course of last session by Mr. Finch. Of fifty-six species of shells observed by Mr. Conrad in this deposit, which extends through Maryland, Virginia, and the county of Cumberland, in New Jersey, one third still exist on the coast of America, but some species in a more southern latitude than that in which they are found fossil.

The Miocene beds, if they occur, have hitherto escaped detection. The Eocene, the middle tertiary of Mr. Conrad, which in England is known as the London clay, and in France as coarse limestone, assumes in America the character of siliceous sand, and in that form has been traced in a north-eastern and south-western direction from Alabama, through South Carolina, Georgia, and Florida, as far as the Gulf of Mexico. Two hundred and nineteen species of shells found in this deposit have been described by Mr. Lea, but among them all, there is perhaps not one entirely analogous to any living species. Several of these shells belong to genera unknown upon the coast of America, some to genera found fossil in Europe, some to genera entirely new. It may be doubted whether any of the species correspond with any of the eocene fossils of Europe, but the number of turreted shells and generic resemblance satisfactorily establish the epoch to which they belong.

It appears from the observations of M. Dufrénoy that in the chalk of the Pyrenees fifty species, in a list of about two hundred, have the character of tertiary shells. A corresponding gradation in the fossil contents of the tertiary and cretaceous formations is

observable in America. The Chalk, or rather the Chalk-Marl, of the
new continent occupies large tracts in New Jersey, Delaware, and
Alabama, and contains among other organic remains teeth of the
Mosasaurus, in no respect differing from those collected at Maestricht.

Mr. Rogers recovers the Chalk formation on the banks of the
Missouri, and about the mouth of the Omawhaw ; its transverse
limit is not known. No flints appear in the beds, but flint nodules,
like the English, occur plentifully lower down the river, even to the
Mississippi.

The Ferruginous Sand of America reposes in the northern states
of the Union as in Sweden and along the Carpathian mountains,
upon primary rocks ; in the southern, upon limestone, perhaps our
mountain limestone ; it forms an irregular crescent, extending
nearly three thousand miles, through Jersey, Delaware, Maryland,
Virginia, the two Carolinas, Georgia, Alabama, Mississippi, Te-
nessee, Louisiana, Arkansas and Missouri.

Dr. Morton and Mr. Rogers refer this formation to the Green-Sand
of England with more confidence perhaps than their observations
warrant. Sands red and green occur in Europe both above the
chalk and below it. Zoological evidence rather militates against their
conclusion. With one or two exceptions all the species are peculiar
to the western continent. *Pecten quinquecostatus*, the only shell
which is quoted as being common to the sands of the United States
and this country, occurs also at Maestricht, and Baculites are characteristic of the upper part of the chalk. From the occurrence of
great quantities of lignite in this formation, from the remains discovered in it of the Scolopas, a bird which inhabits the sea-shore,
and from the locality of the beds in reference to the ancient coast
line, Mr. Rogers infers that the deposit took place in shallow water,
along a coast, which like the present, presented a very extensive
range of soundings ; to this circumstance he attributes the difference of the American and European species of sea shells at the
same period.

With greater probability, as far as the evidence of fossils is concerned, Mr. Rogers attributes to the Green-Sand Formation of England a deposit traced from below the Big Bend to the Rocky Mountains both on the Missouri and the Yellow River, characterized by
Hamites, *Gryphæa Columba*, and *Belemnites compressus.* Above the
Big Bend horizontal beds of lignite, sandstone, shale and clay, occur
continuously for several days journey. The author considers this
formation to be of more recent birth ; it contains, near the Cherry
River, beds of lignite from three to nine feet in thickness.

The New Red Sandstone, with its usual accompaniments of sand
and gypsum, appears to be in North America developed very extensively. According to Mr. Rogers, it comprehends all the country from the falls of the Platte to the great salt lake, or rather sea,
on the western side of the Rocky Mountains, and from the Missouri
to the Arkansas and Rio Colorado. The same formation is supposed to extend into Mexico, and to be the red sandstone described

17

by Humboldt as occurring so extensively in the southern provinces. On ascending the Missouri from its junction with the Mississippi the cliffs are found to consist of Limestone, characterized by Productæ, Terebratulæ and Encrini. The hills near Cheriton are composed of this limestone, and good beds of bituminous coal occur in the same district.

The relative position of the vast deposits of Coal and Anthracite which have been discovered in America is not yet satisfactorily ascertained. The great coal-field of Pennsylvania is said to occur in the higher beds of grauwacke, but what are so called may possibly be shown hereafter to correspond to the limestone shale and millstone grit of Derbyshire. When skilfully treated, this anthracite is considered better than the best bituminous coal of England and the United States. Vegetable impressions are rare, and I do not find that any of the Species have been identified with the English, but the Genera, I believe, are the same. The next great deposit of anthracite, that of Rhode Island, lies rather lower in the series, and the anthracite of Worcester is said to occur in an imperfect mica slate, associated with gneiss. Dr. Meade states, that at Rhode Island the veins of coal are separated by various coloured sandstones of the transition series, yet fine specimens of indurated talc and green asbestus in capillary crystals are also interspersed through the shale, and form the immediate cover of the coal.

The Rocky Mountains, as far as Mr. Rogers could collect from the information of Mr. Sublette, a person engaged for eleven years in the fur trade, and from the journals of Long and Lewis and Clarke and Nuttal, are Primitive. The eastern chain, called the Black Hills, consists of gneiss, mica slate, and greenstone, with amygdaloid and other volcanic substances. Volcanic mounds are frequently seen on the west of the mountains between the rivers Salmon and Louis; for the distance of more than forty miles the Columbia river flows between perpendicular cliffs, from two to three hundred feet in height, composed of lava and obsidian. The Malador branch of the Columbia takes its direction through a similar gorge, and thermal springs abound in this part of the country.

On the various organic remains of North America, a Paper by Dr. Harlan, which first appeared in the Transactions of the Geological Society of Philadelphia, has been republished in Jameson's Journal.

A valuable Communication on the Bermudas, with which we have been favoured by Lieutenant Nelson, R.E., has taught us that in explaining the formation of strata our homage is not exclusively due to Neptune, Vulcan, and Pluto, but that Æolus must also be regarded.

This cluster of islands consists entirely of coral, of what kinds it is unnecessary to specify here, though the author has bestowed upon this part of the subject a large share of attention. Confining myself to what relates more especially to geological science, I may state the following as the most important conclusions which Lieutenant Nelson's observations tend to establish: 1. That the coral animal does not build above water. 2. That coral islands now in process of forming may and do attain a considerable height, say 260 feet above the level of the sea, without the assistance of volcanoes, earthquakes, or any other violent catastrophe. 3. That this height has in Bermuda been attained by a mere accumulation of sand and shells, continually blown up and advancing from the coast into the interior. 4. That drift sand is capable of arranging itself in strata. 5. That of the strata so formed some may be consolidated, others unconsolidated, and that the two may alternate. 6. That strata of drifted sand do not present horizontal surfaces. 7. That wind is capable of giving to strata the figure of a dome or saddle, or a waved and contorted appearance, or an arrangement round centres, or a high degree of inclination. 8. That in coral islands bays are original indentations, not the effects of subsequent abrasion. 9. That the surface of a country may be diversified by hill and dale, though it has never undergone diluvial action. 10. That under favourable circumstances denudation may be occasioned by wind as well as by water. 11. That the ripple-mark, which Mr. Scrope* ascribes to a vibratory movement of the lower stratum of water, agitated by winds or currents, may also be owing to wind. 12. That crevices or fissures may be the results of contraction or unequal expansion, and are not necessarily accompanied by violence. 13. That the reticulation of such crevices does not disprove their being contemporaneous. 14. That caves may be produced in strata by the undermining action of the sea. 15. That limestone may be consolidated without the application of either heat or pressure.

The Bermuda Islands furnish a striking example of the intermixture of land and sea shells with the bones of birds and tortoises, and likewise with vegetable remains. Some of the specimens which accompany the paper have a structure distinctly oolitic, and in some I observe the delicate red tint which is met with in the chalk beds of Yorkshire, or the oolite of Dijon. The cause of this, and still more, the origin of the sand, the detritus of rubies which occurs in one part of the shore, are curious subjects of inquiry. It is also remarkable that breccias should be found at Bermuda, similar to those of Nice, the island of Cerigo, and Gibraltar.

A paper on the arrangement of Fossil Fishes, read at the first meeting after the recess, and ably commented upon by its author, M. Agassiz, received from you more than usual marks of approbation. M. Agassiz informed us, that as yet he had not found any species identical with those of our present seas, with the exception of one small fish which has been discovered in Greenland imbedded in geodes of clay, the geological age of which is undetermined. In the newer tertiary formations, viz. the Crag and superior Apennine beds, the species for the most part exhibit a relation to the genera which dwell within the tropics, but in the older tertiary, viz the London clay, the marine limestone of Paris and the rock of Monte Bolca, at least a third of the shells belong to genera that are ex-

* Proceedings of the Geological Society, No. xxi. 1831.

tinct. In the Chalk more than two thirds belong to extinct genera, and if the grouping of strata were regulated only by ichthyological considerations, this rock would be more properly classed with the tertiary formations than the secondary. Below the Chalk not one recent species has been met with ; in the Wealden Beds, the Oolitic Beds, and the Lias, even the genera are all different from those in the chalk. Below the lias, two out of the four orders, under which M. Agassiz comprehends all the fishes that are known, viz. the Cycloidean and the Ctenoidean, entirely vanish, while the other two orders, rare in our days, suddenly appear in great numbers, together with large sauroid and carnivorous fishes. Of the fishes that occupy the Transition Rocks few have been brought to light, and no peculiar character has yet been affixed to them. In general the more ancient fishes are the best protected by scales. Those which are more ancient than the green sand exhibit none of those marks by which we can determine in the fishes of our own times whether the water in which they live be fresh or salt; the species always changes with the formation, and frequently, as we see, the genus also. It would appear, therefore, that greater changes take place in the higher order of animals than the lower in equal periods of time.

Your award of the Wollaston medal to this eminent naturalist has led to the most advantageous results. By that award M.Agassiz having been induced to come over to this country, has received in all quarters that distinction which his superior knowledge and personal character and deportment justly deserve. With a view to enable him to devote a larger portion of time to the study of fossil Ichthyology in Great Britain, the Association for the Advancement of Science voted to him at Edinburgh the sum of 100l. During his subsequent excursions in various parts of England and Ireland he had ample opportunities of visiting whatever collections have been made in that department of natural history to which he devotes himself; and every one was happy to transmit to our apartments at his request any specimens which he wished to figure. In the very short space of time to which his stay in this country was necessarily confined, M. Agassiz was enabled to add to the very large number of species which he had already examined, no less than two hundred that were entirely new to him ; these were placed, immediately as they arrived, in the hands of an artist from Neufchatel, acting under M. Agassiz's direction. Such are the facilities and advantages which Associations like ours supply to those whom our motto designates as true sons of science!

Sir Philip Egerton has drawn out for us a Catalogue of a rich Collection of Specimens formed by himself and Lord Cole in the caves of Franconia and the Hartz Mountains. In the cavern at Galenreuth, now closed against visitors, it was their good fortune to obtain several bones of the fossil bear, which the late Baron Cuvier required to complete the skeleton of that animal. Many of them appear to have been scratched, but none gnawed. In all these caverns, recent bones referrible to various animals, accompany the fossil ; and in some of

B

them have been found old coins, iron implements, and fragments of rude pottery.

To a work which I shall have occasion to bring under your notice hereafter, Mr. Broderip, our Vice-President, has appended a Table, showing the situation and depth at which the different genera of shells are found in seas and estuaries. The importance of such a table, though professedly incomplete, must be evident to you all; and I hope we shall receive from the same quarter further proofs of the advantage which our science is capable of receiving by allying itself with practical zoology. One of my predecessors has adverted to various circumstances which may determine different fossils to different localities, producing an abundant supply in one place, and a comparative dearth in another. In this point of view, Mr. Broderip's table will be found of great use. By referring to it we discover at once what genera in the present creation are confined to shallows; what genera are to be expected at depths varying from a few feet to three or four hundred, and even more: which are those that attach themselves to marine plants, drifted wood, coral, crustacea, loose stones, or rocks ; perforate the shells of other animals, coral, wood, or arenaceous and calcareous deposits ; or dwell on beds of mud or sand. A knowledge of the habits of recent Testacea must materially assist our investigations into the habits of corresponding fossil genera.

The Geographical Range of different fossil Animals is a subject of great interest, coinciding as it must do with the range of those conditions which were essential to their birth and preservation. To Col. Sykes we are greatly indebted for bringing under our notice a Collection of fossils made by Captain Smee in the district of Cutch. On comparing these remains with the fossils of the oolitic series in England, we observe, not without surprise, that one agrees in every respect with the *Gryphæa dilatata*, that another agrees equally with the young shell of *Trigonia costata*, that a third bears a close resemblance at least to the *Ammonites Harveyii*, while others are identical with the *Ammonites Wallickii*, which is found in the range of the Himalaya. I hope the interest which these species have excited will lead to a more extended investigation of the tract from which they were procured.

Among the illustrations which will accompany a Geological Account of Gurhwal and Sirmou, drawn up by Mr. Royle, is a plate representing certain Shells collected by Mr. Gerard in the elevated valley of the Spiti; these shells may also be identified generically with those of the secondary formations in Europe. Besides these, are given the head and teeth of a small species of Deer, and the tooth of a Rhinoceros, obtained by Messrs. Webb and Trail from the lofty region on the north of the Snowy Mountains ; and several teeth of a carnivorous animal, a saurian, and fish, discovered by Mr. Cauntly at the southern base of the Himalayan chain.

A list of Fossils found in the Lias of Yorkshire by Mr. Williamson, jun. of Scarborough, will be valuable to us as a local monograph, and still more so as a type to which we may refer the fossils of the same formation in other parts of the country. The author is of opinion that every particular layer is characterized by peculiar organic remains, a statement which must be received with caution; it may be true where the district examined is very small, but published lists, which deserve the greatest confidence, establish beyond all doubt that those species which abound in a formation belong to various beds, and that those species which at one locality are most numerous fail altogether in another.

Mr. Mammatt, one of our Fellows, has embodied the result of forty years experience in a splendid Work on the Coal-field of Ashby de la Zouch, illustrated with beautiful engravings.

It is a work of considerable labour, and independently of its local interest, contains some remarks on fissures, joints, and "slines," which coming from a practical observer are well entitled to attention. It is to be regretted that the specimens from which the drawings were taken have been too frequently imperfect or indistinct. In other branches of natural history, so intimate are the relations of the several parts, that from the examination of an unknown tooth and a few other bones, the expert physiologist has been enabled in some instances to construct in imagination an unknown animal, the fidelity of which, to nature, subsequent discoveries have established. The laws which determine vegetable forms are more indefinite and obscure; if fragments of plants are to be engraved, they require to be selected with great judgement, and should be confined to those parts of the object which are best defined and most characteristic. Seen from another point of view, however, these plates become immediately valuable; for though the objects engraved were too indistinct perhaps to enable us to determine the genus, class, or order to which they belong, a correct delineation of them may be sufficient to enable us to identify them with objects found in other coal-fields, perhaps in very distant parts of the world.

The author adopted at an early period the opinion that "Strata are characterized by their Fossils," and he appears to think, that in the coal-field under his consideration, each bed of shale has vegetable impressions of its own. By the precision with which the work is executed, the justice of this opinion is at once put to the test; the successive strata are numbered in regular order, and the names of the plants (where they have names) are attached to the numbers to which they respectively belong. Now, in looking over the list, with a view to the determination of the question before us, I observe *Stigmaria fucoides* at No. 25, 55, 223, 232; *Sigillaria Organum* at No. 37 and 118; *Sigillaria oculata* at No. 74 and 79; *Asterophyllites longifolia* at No. 16, 330, and 370; and *Neuropteris gigantea* at No. 112, 147, 249 a, and 406.

A series of Vegetable Impressions transmitted to us by Mr. De la Beche has given rise to a good deal of discussion. The plants have been examined by Mr. Lindley, and identified at once with those usually found in the Newcastle and other regular coal-fields; they form the roof of certain beds of coal or culm which have long since been observed and worked in the neighbourhood of Bideford in Devonshire, and extend from the shore inland to the distance of about fourteen miles, being about three quarters of a mile in breadth. Along the coast very distinct sections are exposed, both of these beds and their associates. The associated beds have hitherto been generally referred, and with the utmost confidence to the transition epoch. Many of them appear to me identical with those in the Hartz Mountains, to which the name of grauwacke was in the first instance applied, and which may therefore be considered as the types of that formation. Mr. Smith, indeed, in his geological map of England, refers the Bideford district to the red and dunstone of Brecon and the south-east part of Scotland; whether he applies that term to the old red sandstone exclusively, or to the old red sandstone and grauwacke conjointly, I do not know; at all events, neither he nor any other person has ever expressed a suspicion that the beds under our consideration may be more modern than the limestone of Derbyshire; nor am I aware that such suspicion is entertained even now by any one who has seen them in situ.

Mr. Ainsworth, an active naturalist, who is gone out with Captain Chesney on an expedition to the Euphrates, has published an Account of certain Caves at Ballibunnian on the coast of Kerry. In the bay which bears that name, the cliffs which rise to the height of a hundred feet, are composed of two beds (varying from thirty to forty feet in thickness) of compact ampelite, divided by seams of the same slate but fissile and anthracitous, and pouring out streamlets of water containing iron and salts in solution. Near Hunter's Path are seven beds of anthracite; the laminar and slaty rocks belonging to the great transition clay-slates repose on compact sonorous argillaceous limestone, and considerable beds of quartz occur in the midst of the slate formation; this coast, therefore, seems to be very analogous to that of Bideford: it is desirable to ascertain whether the beds of anthracite are here also accompanied by impressions of plants, and whether they can be identified with those of the independent coal districts.

The distinctness of the Bideford section, and the great experience which Mr. De la Beche possesses in geological surveying, make it highly improbable, I think, that the plants which he has presented to us can belong to any other formation than that to which he has referred them: that the same fossils, vegetable as well as animal, are confined to one particular epoch, and cannot be found in more than one part of the general series, are presumptions, which if countenanced, as to a certain extent they are, by limited experience, more enlarged experience may not unnaturally be expected to overthrow, unless indeed we choose to suppose, amid all the obscurity that surrounds us, that our knowledge has already reached a maximum, and that nothing more can ever be visible than that which we have been accustomed to see; but the case which Mr. De la Beche has stated is not altogether a new case; it does not even contradict our present experience. Coal measures, with their usual plants, have been before found in undoubted grauwacke at the

Bocage in Calvados, and you have heard, from one of my predecessors, that they occur in the same relative position at Magdeburg; that they occur in sandstone beds that alternate with mountain limestone in our own country; and that on the southern flank of the Alps they had been discovered by M. Elie de Beaumont in beds of the age of lias.

Two Communications have been presented to us, one from the pen of Mr. Babbage, the other of Captain Basil Hall, R.N., on the Temple of Serapis at Puzzuoli, one of the most extraordinary buildings in Europe; beautiful as a work of art, interesting as an object of antiquity, but to the geologist more especially valuable, as exhibiting a variety of complex natural phænomena, which, though they have taken place in times comparatively modern, it is exceedingly difficult to explain according to the known laws of nature.

Of the solutions which have been proposed by different authors*, not fewer than twenty in number, and most of them authors of eminence, it is impossible to give even a summary within the time allowed me. The merit of Mr. Babbage's paper, as far as original observation is concerned, consists principally in his notice of various stalactitical deposits, and his examination of their different characters and modes of production.

Mr. Babbage describes in detail all the appearances of this temple, and then inquires into the causes of the extraordinary revolutions which it must be admitted on all hands to have undergone : the principal difficulty, you are aware, is to account for the erosion of the columns by lithophagous animals, from the height of 11 feet to 19 above their base, the remaining parts being intact.

Mr. Babbage is of opinion that the building stood at first very nearly at its present level. Assuming that since that period it has both subsided and risen again, and that considerable changes have taken place in the relative levels of the land and sea in its vicinity ; he explains these circumstances by supposing the edifice to have been built upon the surface of matter at a high temperature, which matter contracted afterwards by slow cooling; that at a still subsequent period a fresh accession of heat produced a new expansion, and that in this way the temple was gradually restored to its original level.

To suppose and illustrate his reasoning, the author has constructed a Table (founded on experiments made in America,) showing in feet and decimals what would be the amount of expansion in beds of granite from 1 to 500 miles thick at various temperatures; together with a formula for calculating the amount of expansion in similar volumes of marble and sandstone ; this Table has a collateral claim to notice, as being the first worked out by the calculating engine with a view to publication.

---

* Among these may be mentioned Barthelemy, Boué, Brieslak, Brocchi, Cochin, Billiard, Daubeny, Desmarest, Desnoyers, Forbes, Goethe, Hoff, De Jorio, Lyell, Pini, Prevost, Nicolini, Raspe, and Dr. Robertson.

It appears to me, that in applying the calculation, it is very necessary to take into account three elements which have been overlooked.

1. How far under the supposed conditions expansion would be counteracted by pressure?

2. What space of time would be required to heat or cool such enormous masses of substances, which are very imperfect conductors of caloric, to the required temperatures*?

3. How far the explanation given of the phænomena of the Serapeum is applicable to others in its vicinity? The admitted fact that certain buildings which have also subsided still remain below the level of the sea, while others have been raised to unequal heights above it, makes it unlikely that any uniform cause, while it produced upon them such various effects, should yet have stationed the pavement of this temple in the self-same spot which it occupied at the time of its original construction.

The Letter from Capt. Basil Hall contains remarks on the position of the three columns of the temple which are still standing ; they appear from his observations not to be exactly upright, but to bulge over a little, all in one direction, but not to the same degree. The outermost, in consequence of the tilt, has been brought into such a position that the top of the column is in a line with its base, an extraordinary accident. These remarks do not diminish our difficulties. The tilting may have been occasioned by such subsidences as all buildings are liable to, which are not founded upon solid rock, or it may be referred to earthquakes or original carelessness, or to the skill of the architect, who, by giving a deviation from the plummet line to the axis of the columns (so slight indeed as not to catch the eye,) strengthened his edifice against some lateral thrust, a practice known to have been employed at Athens, and referred to in the letter. Captain Hall has indeed put his veto on this last hypothesis, by saying that the inclination of the columns takes the opposite direction to that which would be required for the supposed purpose ; but this cannot be known, I imagine, unless we know also the details of the original structure, and especially the position and construction of the roofs.

Shortly after this temple had been examined by Mr. Babbage, an attempt to drain it effectually was made by the Neapolitan Government : the stagnant water which infected the air of the neighbourhood was partly supplied from the mineral spring, partly from rain, partly from the sea: the experiment failed for reasons which it is

---

* On a statement of Mr. Scrope's, that a current of lava after it had been ejected nine months, was still flowing on the flanks of Etna at the rate of a yard per day, Mr. de la Beche observes, "If lava can retain its elevated " temperature when thus exposed, what length of time may we not allow for " its doing so within the pipe of the volcano itself, surrounded on all sides by " matter greatly heated, and like itself an exceedingly bad conductor of heat? " Even in those cases where centuries elapse between the great eruptions " of any given volcano, the lava is probably liquid beneath at no very con- " siderable depth."

not necessary to mention. Signor Nicolini, President of the Royal Academy of Naples, who was entrusted with the conduct of the work, and has published an account of it, discovered a rich mosaic pavement a hundred palms in length at the depth of sixteen palms below the level of the stagnant water, whence it appears that the sea must have risen sixteen palms since that pavement was laid. The existence of two pavements of different dates shows further that great changes of level took place before the present temple was built: but Signor Nicolini goes further, he advances a confident opinion that the level of the Mediterranean, in relation to the land, is even now changing.

In support of this doctrine, he not only refers to the phænomena of the temple of Serapis, but points out others in its neighbourhood, all tending to the same conclusion: he states, that in the year 1808 he spent ten days or more in sketching at this spot, and never once saw the pavement flooded, whereas during the last five years he has never once found it dry; that in 1790 the old road near the Serapeum being subject to be flooded, a new one was made at a higher level; and that at the commencement of the new road there is now visible, two palms below the sea level, the pavement of an old landing-place; that boats now pass near the promontory of Puzzoli over a mass of tufa, which derived its name of " The Table," from having been formerly used as such by sea-faring people ; that the ground floor of the Aspizio dei Capuccini is now under water ; and that near Pizzo Falcone modern marks are seen on the tufa many inches under the level of the sea at low water.

Before I quit this branch of the subject, I would wish to suggest to future visitors of this temple, the following topics of inquiry.

What parts of the building have undergone repair? Can the date of these repairs be deduced from the nature of the materials employed, or the character of the workmanship?

Where is the pavement out of level, and to what amount? Are the subsided parts under the lines of thoroughfare, or can their sinking be traced to imperfect construction? Is the foundation such as an architect would call secure? Does it stand on stratum No. 6 of Mr. Babbage's section?

Were there roofs to the bath-rooms?

Would the fragments No. 6, 7, 8, form one column, or more than one ? Was the original number of large (*cipollino*) columns greater than four?

Is the tufaceous deposit on No. 7 the same as that on the walls?

Are all the water lines horizontal?

Brick-work is found in the strata which buried the temple. What is the character of this brick-work? Is it reticulated?

Draw up a detailed and exact account of the strata.

What is the nature of the thermal spring? Evaporate a few gallons of the water, and send the deposit to the Society.

The plan which accompanies Mr. Babbage's paper being copied from that of Jorio, it is desirable, in order to prevent confusion and save expense, that this plan, with the numbers attached to it, should be adopted in any future description.

In the concluding part of the paper, Mr. Babbage proceeds to show in what manner existing causes may possibly elevate continents and mountain ranges, and a similar train of reasoning seems to have presented itself to Mr. De la Beche's mind about the same time. The justice, or at least the relevancy of the reasoning, depends on the establishment of many postulates which in the present state of our knowledge can be regarded only as matters of surmise: but I treated this subject so much at large on a former occasion, that I will not detain you with any further observations upon it now.

A Paper by Dr. Turner, our Secretary, informs us of some experiments which have been made on the action of high-pressure steam upon glass, and other siliceous compounds. The glass was suspended within the boiler of a steam-engine, encased in wire gauze at a temperature of about 300° commonly for ten hours a day. At the end of four months all the pieces were decomposed, and the plate-glass especially, consisting of silex and soda, was in some pieces corroded entirely through. Window-glass was less acted upon, and rock crystal wholly unaltered. Dr. Turner ascribes these changes to the influence of the water on the alkali of the glass, the white opaque matter with which the decomposed pieces were coated being siliceous earth entirely free from alkali; but some portions of the silex also must have been dissolved, for the apertures of the gauze were in some instances closed by a siliceous incrustation, and a small stalactite of silica was found depending from the lowest part. He points out the bearing of these results on the agency of water under high pressure on felspathic and other rocks containing alkalies, and in this point of view they are of great interest.

I hail with unfeigned pleasure the arrival of every paper which makes geology a science not merely of observation, but experiment. In the condition in which we stand at present, the geology of the laboratory is as essential to our progress, as that of the open air.

The Metamorphoses of rocks which are so continually pressed upon our notice are capable of explanation by chemists only. Of those metamorphoses I will only observe, that they appear to me to be attributed too exclusively to Plutonic action. The phænomena which startle and delight us in the vicinity of whin dykes, we regard in Neptunian rocks without emotion.

The Account recently published by Professor Hoffmann of the marble of Carrara is very striking. The result of his examination is, that this pure saccharine limestone in which no trace has been discovered of organic matter, although in its cavities are occasionally found pellucid crystals of quartz, is only transformed oolite. Mr. De la Beche's researches along the gulf of Spezia, an account of which is published in the Transactions of the Geological Society of France, had already prepared us for such an announcement: yet it seems strange when we reflect on the wide expanse of serpentine which is seen in its neighbourhood, that the Carrara marble should not be magnesian. In the Isle of Skye veins of serpentine sometimes penetrate the lias, where, in the vicinity of numerous whin dykes it assumes the whiteness and occasionally the sparkling grain of statuary marble, and here again the marble is unadulterated by magnesia : the origin of the serpentine is somewhat less mysterious since the limestone in its unaltered state is micaceous. M. Dufrénoy in a late number of the " Annales des Mines," has described a similar Transformation of lias into saccharoid limestone seen in the Pyrenees. I think it unnecessary to detail to you the descriptions which Mr. Murchison has given of the Change of structure, or even of substance that take place at the Malvern Hills and at sundry other places in Wales, and on the confines of Wales frequently, though not always, in the vicinity of trap and sienite, because they are in general the same as have been observed repeatedly in other districts. The phænomena at Old Radnor the author remarks are very analogous to those of the Val di Fassa in the Tyrol. It may, however, be proper to mention, not as a novelty, but as a circumstance the frequent occurrence of which is little attended to, that in Carmarthenshire the line of altered rock produced by the proximity, or as it is called the protrusion of a mass of porphyritic trap, is parallel to the strike of the grauwacke so altered. At Caer Caradoc, the Wrekin, the Stiperstones, and elsewhere, a stratified sandstone is at its junction with trap, converted into quartz rock. One other circumstance deserves to be noticed; the range of the Stiperstones, along which these Plutonic appearances present themselves, is flanked on either side by metalliferous deposits, but not of the same kind, the copper ores being all found on one side of the range, the lead ores on the other. An analogous case will be seen in Humboldt's account of the country situate between the Oural and Altaic chains. A fault or fissure is there traceable through not less than 16° of longitude, forming a crest or water-shed; the rocks are nearly the same as those of Shropshire : they comprehend a granite (unconnected with gneiss,) clay-slate, grauwacke-slate, augitic porphyry, and transition limestone, once compact but now granular. Malachite and red Copper ore are found on one side of the ridge, argentiferous galena on both.

Such are the Plutonic phænomena, for an explanation of which we rely chiefly on the assistance of chemistry; but there is another train of phænomena which renders a close and intimate alliance between this science and our own, no less desirable. The spontaneous generation, shall I call it? of agate, of chert, of hornstone, of flint, in the midst of sedimentary calcareous deposits, apparently through the instrumentality of animal or vegetable matter, in which little or no silex is to be met with, is one of those mysterious operations of nature which can nowhere be satisfactorily accounted for unless in the laboratory. The coralline agates of Antigua, the entrochal cherts of Derbyshire, the siliceous shells of Blackdown or Fontainebleau, the chalcedonic alcyonia of Pewsey, pieces of fossil wood either imbedded in strata or loosely scattered over sandy deserts, the flinty casts of echini and other substances in the midst of our chalk, all these suggest a course of experimental investigation which we are entitled to hope, if undertaken in earnest, would not be undertaken in vain.

Gentlemen, I have great satisfaction in announcing to you, that at the opening of the present Session of the Royal Society, one of the Royal Medals was awarded to our Foreign Secretarv as the author of the most important discoveries or series of investigations sufficiently established or completed to the satisfaction of the Council within the last five years, and for which no honorary reward had been previously received. The Council of the Royal Society, premising that they decline to express any opinion on the controverted positions contained in Mr. Lyell's work, entitled " Principles of Geology," state the following as the grounds of their award.

1. The comprehensive view which the author has taken of his subject, and the philosophical spirit and dignity with which he has treated it.

2. The important service he has rendered to science by especially directing the attention of geologists to effects produced by existing causes.

3. His admirable description of many tertiary deposits, several of these descriptions being drawn from original observations.

Lastly, The new mode of investigating tertiary deposits, which his labours have greatly contributed to introduce ; namely, that of determining the relative proportions of extinct and still existing species, with a view to discover the relative ages of distant and unconnected tertiary deposits.

Of the Work so honoured by the Royal Society, the third edition has been lately published: in this edition some opinions formerly expressed have been modified or renounced, and much new matter has been introduced ; the phænomena of springs and Artesian wells have been more fully treated ; the theory of elevation has been entered into more at large, the author still controverting that theory. A chapter, almost entirely new, points out the probable causes of volcanic heat ; objections are advanced against the doctrine of the central fluidity of the earth, and especially the intense heat attributed by some writers to the internal nucleus. Mr. Lyell considers how far chemical processes in the interior of the earth may generate volcanic heat, and what may be the effect exerted by currents of electricity. Sir Humphry Davy's theory of an unoxidated metallic nucleus is considered, and it is suggested that compounds resulting from the action of water upon metallic bases may be again deoxidated by the hydrogen set free in that process. The author has also given a more complete view of his opinions respecting the origin of caverns, and the manner in which they have been filled with breccia and the bones of animals. In illustrating this subject, he refers particularly to the recent discoveries of MM. Virlet and Boblaye in the Morea. His sketch of the principal secondary formations is also considerably enlarged.

Two other publications have issued from the press during the last year, which are eminently deserving of your attention. The first of these, entitled "A Treatise on Primary Geology," originated in great measure from a discussion that took place at a Meeting of the Geological Section of the British Association at Cambridge, and was drawn up with a view to the further consideration of the chief questions which it embraces, at the subsequent Meeting of the same body at Edinburgh.

Dr. Boase begins by describing the composition and relation of the several Primary Rocks, combining the accounts of geologists in various parts of the world with the results of his own laborious investigations in Cornwall. I regret, that within the limits to which I am restricted, it is impossible for me to do justice to the merits of this important work ; I must confine my observations to a few of its most characteristic features. Dr. Boase is of opinion that the connexion between Unstratified and Stratified Rocks shows that they had a common and contemporaneous origin. He observes that granitic masses are as complex in their composition as stratified rocks, and form sometimes distinct regular beds, highly inclined and alternating with one another; that the elvans, or insulated beds of granitic rock, always partake of the nature of the containing slate, and have the same basis ; that the difference between the granite and killas, or elvan, in Cornwall is often feebly marked, and still more feebly the difference between the granite and gneiss of Scotland and Germany; as little difference is there between the granite of the Alps and the talc slate adjoining. Where the granite changes its character, a corresponding change, he says, takes place in the slate. The elvans are connected on either side with the granite they intersect by the most intimate mineral gradations, or contain irregular portions or masses of the common granite, with which they coalesce ; both are penetrated by crystals of felspar; both are striped with shorl. At Pednimerer Meer one of the parallel layers of granite runs through the elvan. In Scotland different beds of granite will intersect a common mass, and pass by minute mineral transition into one another, or into the characteristic granite of the district.

Dr. Boase considers that the Primitive Shisti are improperly said to be *stratified*. Pini has expressed the same opinion in two separate memoirs; the supposed planes of stratification are, in his view of the subject, mere transverse fissures. Prof. Phillips, Mr. Scrope, Dr. Fleming, Prof. Sedgwick, have all felt and expressed the difficulty of distinguishing in these shisti planes of cleavage, and planes of stratification. In the days of my geological apprenticeship I took great pains to observe and record dip and direction, and fondly hoped to obtain so large a number of accurate data on this subject as might enable me to arrive at last at some general and important result. I threw these data into tables, which only bewildered me. Suspecting the accuracy of my early observations, I repeated them again and again, guarding myself on every occasion more and more against probable sources of error; still I found my results perpetually varying, till at last my patience was exhausted, my Clinometer discarded, my registers destroyed. Let it not be

supposed, however, that my observations were useless; they taught
me a salutary distrust.

Dr. Boase remarks that all the Slate rocks are composed of rhomboidal concretions, which are developed on a large scale by disintegration. Mr. Scrope had anticipated this remark, and generalizes it.
He says the stratification of rocks of all kinds, where the strata are
separated by seams, is produced by concretionary process.

Now, then, which of all the planes are the planes of dip? Dr.
Boase, like the Woodwardian Professor, selects those which run
with the laminæ, and yet the layers of massive crystalline and
granitic rocks often lie the other way. But this seems to be very
much a matter of taste; different observers selecting for the scene
of their measurements different planes. Some pay great attention to
the laminæ, others neglect them; nay, the same observer shall sometimes select as strata one series of planes, sometimes another.

Professor Phillips, in a passage which is too long to be quoted,
has expressed the same idea in language equally expressive.

Dr. Boase presents to those who differ from him on this subject
the following alternative: either Stratification implies a successive
deposition of sedimentary matters held in suspension, in which case
none of the primary shisti are stratified; or merely parallel planes
without regard to the cause of their production, in which case *not
only the primary shisti are stratified but granite also.*

In the thirteenth chapter will be found some excellent observations
on the nature of Inclined Strata, tending to show that these last are
not necessarily the effect of disturbance, but are to be attributed, in
the Primary Slates, to original structure, and in many of the Secondary, partly to this cause, and partly to deposition upon inclined surfaces.

The difficulty I have been considering is by no means confined to
Primary Slates. Mr. Conybeare has observed on the coast of Sully,
in Glamorganshire, that the Lias splits spontaneously into blocks of
regular figure, corresponding to that of a crystal of calcareous spar.
If this be the case, where are we to look for the seams of Stratification? I have felt for very many years, and I still feel that the
indistinctness of this term is one of the most dangerous stumbling
blocks we have to encounter. If we would found upon this distinction the grand classification of rocks into Neptunian and Plutonic,
surely we ought to have some test by which to determine whether
rocks are *stratified or not.* If, looking to the theory of M. Elie de
Beaumont, we would know whether strata are conformable or disturbed, surely we ought to be placed in a condition to determine
what *Strata* are. On taking leave, as I must do, of Dr. Boase's work,
I again recommend it to your attentive perusal; it is written with
great candour as well as earnestness, and will be found a useful cor-
rective of many opinions which appear, to me at least, to have been
too inconsiderately adopted.

Mr. De la Beche, one of your Vice-Presidents, to whose pen and
pencil our science has been for a series of years continually and largely

indebted, has published a small volume, entitled "Researches in Theoretical Geology" The main tendency of this volume is to establish the importance and practicability of subjecting geological opinions to the tests of chemistry and natural philosophy. The Author goes over much ground, keeping always the same direction, having apparently no other objects in view than the acquisition and communication of sound knowledge, the detection and exposure of error, and the discovery and establishment of truth. Unshackled by authority, unenslaved by preconceived opinions, unseduced by the love of novelty, free from all vanity of authorship, concise, methodical, exercising his judgement continually, his fancy seldom, the author may not obtain that popularity which with less merit he might have easily commanded; but such a work cannot fail to be appreciated here.

After taking a general view of the Solar System, and considering certain apparent agreements and disagreements in the condition of some of the Planets, Mr. de la Beche applies his observations entirely to the Earth, which he supposes to have been originally in such a state that its component particles had a free passage among one another. The principal Constituents of Land, Water, and Air, sixteen in number are made up of Substances commonly termed simple: viz. oxygen, hydrogen, nitrogen, carbon, sulphur, chlorine, fluorine, phosphorus, silicium, aluminum, potassium, sodium, magnesium, calcium, iron and manganese. Adopting Laplace's hypothesis, that the sun, planets, and their satellites, have resulted from the Condensation of gaseous matter, he ascribes the Condensation of our own planet to the gradual Radiation of Heat into space. He shows how Sedimentary Rocks may be deposited so as to present, from the first, inclined planes, and that we should therefore be cautious in referring to subsequent displacement all deviations from a horizontal level: he forms an estimate of the Destruction of Land by Breakers, of the wear and tear of Running Waters, of the transport of detritus by Rivers.

The mean Density of the crust of the earth is usually reckoned at 2·5. From a reference to the lists, which the author has drawn out, of the specific gravities of many rocks, of the various simple minerals which enter into their composition, and of certain recent shells, it would seem that 2·6 would be nearer the truth.

The Author investigates the Chemical Changes which Rocks undergo after their formation, and the subject of Concretions, such as *Ludus Helmontii*, &c. He remarks on the importance of attending to the Cleavage of Rocks, whether igneous or aqueous, and their Transformation: he considers the great Breaks of the Surface in reference to the effect which would result from its gradual cooling; and, from the contortions and Fractures of Mountain Chains, infers the Intensity of the forces that have acted upon them: he shows that certain Thermal Springs may be occasioned by the Condensation of volcanic discharges of gas and vapour, and ascribes the Uniformity of their Temperature to the Constancy of such condensation: he then treats of Volcanic Action and the gradual Rise of large tracts of Land.

When explaining the Formation of Valleys, Mr. De la Beche contends that the "Bursting of Lakes," as it has been termed, could not take place in the way supposed. The Area, comprised within sound-ings, that is, within the 100 fathom line, round the British islands, is delineated on a map, in order to show, that within that area at least, no Valleys are produced by Tides and Currents ; whence it would fol-low, that such effects cannot be satisfactorily referred to such causes. Under the head of Faults, which are treated of at some length, the author shows with what facility " Craters of Elevation" may be formed, and expresses surprise that so much discussion should have taken place on so simple a case ; he sees no difference between many Me-talliferous Lodes and Faults.

The subject of Organic Remains is next investigated, and the Modi-fication produced by various physical causes on the distribution of Life, particularly Animal Life in the sea. Diagrams are given to show that Shells, contemporaneously enveloped by rocks now forming, would probably not be of the same Species, even under the same parallels of latitude ; but that the Species would be determined in some measure by the relative depth of the water at different places, on the influx of rivers, and other causes. Attention is particularly called to the man-ner in which Organic Remains may be entombed in a series of deposits along a gradually rising coast.

Under the head of Mineralization of Organic Remains, the Author shows that these bodies are not merely changed in character but in reality. One substance being substituted for another, a cast for an original, the Change varies in these bodies in proportion to their re-spective Solubility. Carbonate of lime being more soluble than phos-phate, shells change much more rapidly than bones. Silica in shells follows the same law as in agates, entering their cavities by infiltra-tion.

The Author now gives a general Sketch of the various Rocks that are known to us; he remarks that the Primary differ chemically from the Secondary, and endeavours to account for the phænomena con-nected with animal and vegetable life, as exhibited in the several for-mations, upon the theory of a gradual Loss of Heat by Radiation. Upon this theory he would explain the Scanty Supply of Limestone in the earliest Rocks. The effect of great heat would be to expel from water the carbonic acid necessary to hold the carbonate of lime in solutions ; and consequently Calcareous matter could not be deposited from heated water. He observes also that the Conditions for an uniform distribution of animal and vegetable life, would be more uniform in a thermal than in our actual Seas, and, therefore, if the Ocean had be-come gradually cooler, we should expect to find, as we do, genera and species more diversified.

The terms, Eocene, Miocene and Pliocene, are objected to, as pre-judging an important question. Unless equal conditions obtained at equal times in all places, the Miocene rocks of one country may be of the same date as the Pliocene of another. The Author closes his remarks on Erratic Blocks by observing, that, like the great contor-tions and dislocations of strata, they teach us, while we duly appre-ciate the continued and more tranquil effects which are daily before our eyes, that we must not dismiss from our consideration Forces of greater Intensity; still bearing in mind, that however great these last-mentioned forces may be when measured by the ideas commonly entertained on such subjects, they are still insignificant when considered with reference to the entire spheroid on the surface of which they act.

Gentlemen, I have now brought to a close the account, which, in conformity with the practice of the Society, I proposed to lay before you, of the labours and achievements of last year. It therefore only remains for me to resign the chair. When I consented to resume the office of President, many of you are aware that a consciousness of the precarious state of my health made me diffident of my powers to discharge that office with becoming energy and effect. The generous support which I have received from the Council and the Fellows at large, has, I am willing to believe, in some degree concealed from your observation several deficiencies, of which I myself have been fully aware, and the Society has continued to flourish. The only merit I claim is, that of having, upon all occasions *endeavoured* to promote your interests; but a brighter prospect now opens upon you, and you will find an ample guarantee for more brilliant anticipations of success in the youth, the spirit, the abilities and the character of my successor.

THE END.

Printed by Richard Taylor, Red Lion Court, Fleet Street.

Printed in the United States
By Bookmasters